T0216576

Local Identities and Politics

The relation between identity and space is strong and generates many conflicts. Most people attach great importance to their local community and its identity. The possibility of change can cause turmoil and become fertile ground for staking new identities. Understanding how these changes can take place is important to the future of community cohesion across the world.

This book gives a detailed analysis of how different stakeholders in two Dutch municipalities use and adapt their identity discourses to deal with changing circumstances, situating this work within a wider international context through global comparisons. The growing spatial interdependence and political pressures for municipal cooperation or amalgamation creates not only threats, but also opportunities for stakeholders in local communities to transform their local identities. By studying how local communities attach to local identities, a new conceptual framework can be formed, informed by lively accounts from residents on the rich and varied use of identity in their communities and their concerns over future developments.

This is valuable reading for students, scholars and researchers working in geography, politics, sociology and cultural studies.

Kees Terlouw is a political geographer at the Department of Human Geography and Spatial Planning at Utrecht University, the Netherlands.

Local Identities and Politics
Negotiating the Old and the New

Kees Terlouw

Routledge
Taylor & Francis Group

LONDON AND NEW YORK

First published 2017 by Routledge

2 Park Square, Milton Park, Abingdon, Oxfordshire OX14 4RN
52 Vanderbilt Avenue, New York, NY 10017

Routledge is an imprint of the Taylor & Francis Group, an informa business

First issued in paperback 2018

Copyright © 2017 Kees Terlouw

The right of Kees Terlouw to be identified as author of this work has been asserted by him in accordance with sections 77 and 78 of the Copyright, Designs and Patents Act 1988.

British Library Cataloguing in Publication Data
A catalogue record for this book is available from the British Library

Library of Congress Cataloging in Publication Data
Names: Terlouw, Kees, author.
Title: Local identities and politics : negotiating the old and the new / Kees Terlouw.
Description: Milton Park, Abingdon, Oxon ; New York, NY : Routledge, 2017. | Includes bibliographical references and index.
Identifiers: LCCN 2016042560| ISBN 9781138209251 (hardback) | ISBN 9781315457536 (ebook)
Subjects: LCSH: Overflakkee (Netherlands)–Social conditions. | Katwijk aan Zee (Netherlands)–Social conditions. | Group identity–Netherlands–Overflakkee. | Group identity–Netherlands–Katwijk aan Zee. | Community life–Netherlands–Overflakkee. | Community life–Netherlands–Katwijk aan Zee. | Spatial behavior–Social aspects–Netherlands–Overflakkee. | Spatial behavior–Social aspects–Netherlands–Katwijk aan Zee. | Politics and culture–Social aspects–Netherlands–Overflakkee. | Politics and culture–Social aspects–Netherlands–Katwijk aan Zee.
Classification: LCC HN513.5 .T47 2017 | DDC 306.9425/39–dc23
LC record available at https://lccn.loc.gov/2016042560

ISBN: 978-1-138-20925-1 (hbk)
ISBN: 978-0-367-13880-6 (pbk)

Typeset in Times New Roman
by Wearset Ltd, Boldon, Tyne and Wear

Contents

Figures

Tables

Acknowledgements

This book is partly based on the 58 interviews Maarten Hogenstijn conducted with members of the local communities in Katwijk and Goeree-Overflakkee. He also analysed the local documents on which the topic list of the interviews was based. Maarten's interview skills and dedication contributed greatly to the realisation of this book. He was also so kind to let me use his pictures (Figures 4.1, 4.2, 4.3, 4.4, 5.1 and 5.2). I am also grateful to all those in Katwijk and Goeree-Overflakkee who took the time to explain to us what the role of local identities was in their community. I also thank all those civil servants of the Dutch Ministry of the Interior who were involved in commissioning and supervising the research on local identities. Of course the usual disclaimer applies and I am solely responsible for the content of this book, which does not necessarily reflect the points of view of the Minister of the Interior and Kingdom Relations.

Margot Stoete of our Kartlab was so kind to draw the maps in Figures 3.1 and 3.2.

As this book is the product of almost a decade of research on the use of local and regional identities I sometimes used texts of my earlier publications, for which the publishers were so kind to give me permission. This book is derived in part from an article, "Territorial changes and changing identities: how spatial identities are used in the up-scaling of local government in the Netherlands", published in *Local Government Studies* on 8 June 2016, available online: www.tandfonline.com/doi/full/10.1080/03003930.2016.1186652. In particular, §1.1, §2.9, §5.1, §5.2, §5.3, §6.2 and §6.3 are partly based on this article. I also thank Martin Boisen and Bouke van Gorp with whom I sometimes collaborate on local identity issues. I wrote with Bouke the article on "Layering spatial identities: the identity discourses of new regions" for *Environment and Planning A*, 2014, volume 46, pages 852–866, published by SAGE, from which I used parts in §2.7 and §6.1. In Chapter 2, most of §2.5 and portions of §2.8 are partly reprinted from *Geoforum*, volume 57, Kees Terlouw, "Iconic site development and legitimating policies: the changing role of water in Dutch identity discourses", pages 30–39, copyright 2014, with permission from Elsevier.

I also wish to thank my department for not distracting me with too much interest in what I was doing, and all those at conferences, meetings, visits and even the many reviewers whose comments motivated and somewhat steered my research over these last few years. I end by expressing my warmest gratitude to my family, who suffered the most from my late working hours. My daughters, developing quite an identity of their own, have acquired a healthy aversion to the geographer's perspective on this subject. My wife, Andrea, also stayed with me through thick and thin in this project. I perhaps discussed spatial identities and politics too much with her, but I am grateful for her listening ear and her gentle ways to prevent me from straying too much in my conceptualisations.

1 Introduction

Local interests are adequately represented in the municipal council. But less and less municipal officials are linked to the local communities. These have in a way no longer feelings for local issues and identities.

It slowly trickles down. When you lead the way, you sometimes look back and in the rear people are unaware of who leads the way. Figuratively speaking that is. But if you communicate plentifully in many different ways, the local media, organise meetings, municipal bulletins, then it reappears (...). A kind of brainwash is necessary. But we are only at the beginning of the Gaussian curve to adjust the identity. This is evolution not revolution. That does not suit Goeree-Overflakkee. You have to do that slowly, in small steps.

These two quotes illustrate the different ways in which local identities are linked to local politics. The first quote illustrates that a declining role of local identity in local affairs can widen the gap between administrators and population, which hampers the effective functioning of local administrations. The second quote illustrates how local identity can be changed to support local political projects, in this case the stimulation of local development through island marketing. These two quotes are from interviews which were part of an explorative study into the role of local identities for local communities commissioned by the Dutch Ministry of the Interior (Terlouw & Hogenstijn 2015). This book builds on this material to explore the different ways in which local identities are used in politics.

The main argument of this book is that local identities are not fixed objects and remnants from the past which only hinder the dynamics of future-oriented policies. Local identities are constantly being shaped and reshaped by different actors, who use local identity discourses not only to hinder but also to promote specific local policies. This book analyses the backgrounds of this differential use of local identities and develops some conceptual tools to clarify the different ways in which local identities are used in local politics. This helps to better understand the different ways in which both old and new elements are used in local, regional and other spatial identity discourses.

Spatial identities have always played an important role in politics. The role of national identities in national and international politics especially has been extensively acknowledged and studied. The rise of nationalism and the decline of localism since the nineteenth century in Western countries were commonly regarded as obverse sides of the same coin. The decline of local identities was seen as a welcome sign of the breaking of the bonds of traditional societies which restrained individuals in their path towards modernity. Individual freedom, democracy, economic development and welfare were linked to the imagined community of the nation, which as a Janus not only looked backwards in history, but also forward to the future. Local identities were seen as a relic of the divisive past now finally being surpassed by more future-oriented modern and integrative nationalisms. All nationalisms stress the homogeneity of the nation and the exclusiveness of the relation between the individual and the nation. They leave no room for other significant collective identities (Smith 1982; Yack 2012, 107; Nairn 1997; Billig 1995). Modern individuals were to identify predominantly with their national community. Local and regional identities did not disappear, but were reformulated as the historical backdrop of the national discourse which focusses more on integration and development away from the old localism characterised by fragmentation and stagnation (Confino 1997; see also Box 6.4).

This conceptualisation of local and regional identities as subordinate to national identity has over the last decades been challenged by discourses on globalisation. Some argue that the combination of anonymous globalisation and individualisation results in an increasing identification with the more familiar local and regional environment. The decline of the nation-state is generally linked with the transfer of powers upwards to the international level and downwards to the local and the regional levels. The changing role of the nation-state is linked to the growing importance of local and regional administrations (Brenner 2004; MacLeod & Jones, 2001; Keating 1998; 2013; Hooghe & Marks, 2003; Rodríguez-Pose & Sandall 2008; Jessop 2000; Paasi 2012).

The decentralisation of tasks from the national level puts additional pressures on municipal administrations. Municipalities therefore on the one hand increasingly cooperate and on the other hand are pressured into amalgamations. Municipalities in many different Western countries increasingly cooperate in new functional regions in order to more efficiently and effectively provide public services. These new regions are however often short-lived, lack clear spatial borders and partially overlap in space with other regions, creating "splintered political geographies" (Brenner 2004, 292) and a "continued institutional and spatial disorder" (ibid., 296) of the rescaling state. On the other hand, these new tasks increase in many countries the pressure on municipalities to amalgamate (Baldersheim & Rose 2010; De Ceuninck *et al.* 2010). This not only affects the organisation of local, regional and national administrations – on which there exists a huge literature – but also changes the role of local and regional identities, which is the topic of this book.

The relation between local identities and politics is hardly studied. Most political and academic attention focusses on the growing role of identities in inter and intra national conflicts. Much has been written about the political struggles for regional autonomy in regions like Scotland, Catalonia and Flanders. The role of spatial identities in local politics has received much less attention. Although it is frequently mentioned in political debates on municipal amalgamation, it is hardly systematically studied by academics. The debates on municipal amalgamation have traditionally focussed on the expected economies of scale. Most academics now agree that it is very difficult to realise these efficiency gains. The diseconomies of scale and the costs of the amalgamation process frequently outweigh the realised economies scale, especially when the amalgamation is forced upon the local administrations and populations (Drew *et al.* 2014). Administrative effectiveness has succeeded economic efficiency as the main argument which is used to legitimate municipal amalgamations. The diminishing role of the central state in policy fields like welfare increases the workload for small municipalities. Creating larger municipalities with more, and more specialised, personnel helps to administer these growing and increasingly complex tasks more effectively. Amalgamated municipalities generate economies of scope, while they increase their strategic capacity. These larger municipalities can deliver more complex services and are assumed to better face future challenges than small municipalities (Aulich *et al.* 2014).

Although amalgamation can strengthen the municipal organisation and its external position towards other administrative levels, it can also weaken its internal position towards its own population. Many studies mention the growing distance between the administration and the population after amalgamation causing a decline in political participation, as for instance indicated by diminishing voter turnout. The incorporation of different local communities in an amalgamated municipality also increases the diversity of interests and preferences. This destabilises the local political system, makes consensus formation more difficult and stimulates a form of municipal politics based on confrontations (Kjaer & Klemmensen 2015; De Ceuninck *et al.* 2010; Aulich *et al.* 2014; Hanes 2015).

The growing administrative effectiveness of amalgamated municipalities is thus at least partly offset by a decline in popular support and participation. This alienation of the local population is frequently linked to the importance of local identities. Many studies mention the feared loss of local identities as an obstacle hindering amalgamation. These studies primarily discuss this as the feared loss of local traditions and culture, but do not analyse the character of these identities and how these are used in the local politics (Baldersheim & Rose 2010; Mecking 2012; van Assche 2005; De Peuter *et al.* 2011; Boudreau & Keil 2001; Tomàs 2012; Spicer 2012; Fortin & Bédard 2003; Lightbody 1999; Keil 2000; 2002; Hulst & van Montfort 2007; Alexander 2013; Rausch 2012; Hanes 2015; Aulich *et al.* 2014).

This book aims to contribute to a better understanding of the different ways in which identities are used in local politics through a detailed analysis of the role of local identities before and after amalgamation in two Dutch municipalities. Its results are also relevant for many other countries, where the changing role of municipalities puts pressures towards amalgamation and cooperation. As a result, the pace and debate on the necessity of municipal amalgamations has increased in the last decade in most Western countries (Baldersheim & Rose 2010, 237, 244).

1.1 Municipal amalgamations in the Netherlands

The Netherlands has a long history of municipal amalgamations. This is generally a gradual process, but in some periods the pace of amalgamations increases. At the beginning of the twentieth century, the physical growth of cities into the surrounding countryside was an important reason for the annexation of rural municipalities. Since the Second World War, the focus has shifted to the amalgamation of peripheral rural municipalities with other rural municipalities. No longer expanding cities, but the expanding central welfare state stimulated municipal amalgamations. The rolling out of new public services to all citizens was linked to the formation of more effective, larger municipalities (Beeckman & van der Bie 2005).

However, in the past decade, no longer is it the rolling out of new state functions to its citizens that puts pressures on especially small rural municipalities to merge, but the rolling back of welfare state provisions. In 2012 the Dutch government formulated far-reaching plans to shift the focus of many welfare arrangements from the central to the municipal level. This makes municipalities responsible for a complex set of welfare arrangements for the youth, chronically ill, elderly and unemployed. These plans will almost double the municipal budgets and shift the political focus of municipalities from the regulation of spatial developments in its territory to the distribution of welfare over its inhabitants. The central government contends that most rural municipalities are too small to deal effectively with these new tasks. To speed up the amalgamation process, they declared that 100,000 inhabitants was the desired minimum size of municipalities, but the affected rural municipalities were more or less free to choose with whom they would merge (BZK 2013). This would affect most of the more than 400 municipalities in the Netherlands, as only 30 of them have more than 100,000 inhabitants. Municipalities are not formally obliged to merge, but have to demonstrate that they have the competent manpower to perform their new tasks. Initially, this can be achieved through the "voluntary" cooperation with other municipalities. Municipal cooperation is however seen by many as a temporary intermediate phase. The Dutch government expects that when municipalities get to know each other better and experience the benefits of a larger organisation, they will after some time merge voluntarily.

This discourse preferring large municipalities for their assumed administrative efficiency, effectiveness and democratic control met with strong local resistance in the affected municipalities. They challenge the official administrative discourse on effectiveness and efficiency of larger municipalities. In general, the legitimating power of professionals and technocrats has been greatly reduced over the past decades. Cultural, softer and more locally embedded knowledge can no longer be successfully dismissed as outdated traditional parochialism (Giddens 1990). Until a few decades ago, the dominant modernisation discourse of growth and improvement was hard to resist by rural communities, but now there is more room for counter-narratives. Local communities increasingly use arguments related to their distinct local identity to oppose municipal amalgamations. They frequently argue that their local identity is too different from that of their neighbours and that mergers would create indistinct anonymous large municipalities, which will weaken their distinct local identities and social cohesion (Terlouw 2014; van Twist *et al.* 2013).

The Dutch central government has difficulties coping with these arguments based on local identities. For them they appear to be outside the realm of a public debate, which should be based on general and rational arguments. They question the importance of local identities. How can local identities be that important when they only appear in the political debate during amalgamations, but do not seem to be important at other moments? The Dutch Ministry of the Interior therefore commissioned a study to explore the importance of local and regional identities for local communities. The cases studied in this book are largely based on this study. The full report is published in Dutch by the Ministry of the Interior and accessible through their website (Terlouw & Hogenstijn 2015). Partly based on the results of this study, the Ministry of the Interior produced a new vision document in 2016 in which they further shifted their policies from municipal amalgamation to municipal cooperation. Now they acknowledge more the negative consequences of the formation of local resistance identities during forced municipal amalgamations. They also recognise the positive possibilities of linking local identities with voluntary regional cooperation, especially in order to promote economic development (Studiegroep Openbaar Bestuur 2016).

References

Alexander, D. (2013). Crossing boundaries: action networks, amalgamation and inter-community trust in a small rural shire. *Local Government Studies*, 39, 463–487.

Aulich, C., G. Sansom & P. McKinlay (2014). A fresh look at municipal consolidation in Australia. *Local Government Studies*, 40, 1–20.

Baldersheim, H. & L.E. Rose (Eds) (2010). *Territorial choice: the politics of boundaries and borders*. New York: Palgrave.

Beeckman, D. & R. van der Bie (2005). Een eeuw gemeentelijke herindelingen. *Bevolkingstrends, statistisch kwartaalblad over de demografie van Nederland*, 53, 63–64.

Billig, M. (1995). *Banal nationalism*. London: Sage.

Boudreau, J. & R. Keil (2001). Seceding from responsibility? Secession movements in Los Angeles. *Urban Studies*, 38, 1701–1731.

Brenner, N. (2004). *New state spaces: urban governance and the rescaling of statehood*. Oxford: Oxford University Press.

BZK (2013). *Bestuur in samenhang*. Den Haag: Ministerie van Binnenlandse Zaken en Koninkrijksrelaties.

Confino, A. (1997). *The nation as a local metaphor*. Chapel Hill: University of North Carolina press.

De Ceuninck, K., H. Reynaert, K. Steyvers & T. Valcke (2010). Municipal amalgamations in the low countries: same problems, different solutions. *Local Government Studies*, 36, 803–822.

De Peuter, B., V. Pattyn, & E. Wayenberg (2011). Territorial reform of local government. *Local Government Studies*, 37, 533–552.

Drew, J., M.A. Kortt & B. Dollery (2014). Did the big stick work? An empirical assessment of scale economies and the Queensland forced amalgamation program. *Local Government Studies*, 42, 1–14.

Fortin, A. & M. Bédard (2003). Citadins et banlieusards. Représentations, pratiques et identités. *Canadian Journal of Urban Research*, 12, 124–142.

Giddens, A. (1990). *The consequences of modernity*. Cambridge: Polity Press.

Hanes, N. (2015). Amalgamation impacts on local public expenditures in Sweden. *Local Government Studies*, 41, 63–77.

Hooghe, I. & G. Marks (2003). Unravelling the central state, but how? *American Political Science Review*, 97, 233–243.

Hulst, R. & A. van Montfort (2007). *Inter-municipal cooperation in Europe*. Dordrecht: Springer.

Jessop, B. (2000). The crisis of the national spacio-temporal fix and the tendential ecological dominance of globalizing capitalism. *International Journal of Urban and Regional Research*, 24, 323–360.

Keating, M. (1998). *The new regionalism in Western Europe: territorial restructuring and political change.* Cheltenham: Edward Elgar.

Keating, M. (2013). *Rescaling the European state: the making of territory and the rise of the meso.* Oxford: Oxford University Press.

Keil, R. (2000). Governance restructuring in Los Angeles and Toronto: amalgamation or secession? *International Journal of Urban and Regional Research*, 24, 758–781.

Keil, R. (2002). "Common-sense" neoliberalism: progressive conservative urbanism in Toronto, Canada. *Antipode*, 34, 578–601.

Kjaer, U. & R. Klemmensen (2015). What are the local political costs of centrally determined reforms of local government? *Local Government Studies*, 41, 100–118.

Lightbody, J. (1999). Canada's seraglio cities: political barriers to regional governance. *Canadian Journal of Sociology*, 24, 175–191.

MacLeod, G. & M. Jones (2001). Renewing the geography of regions. *Environment and Planning D*, 19, 669–695.

Mecking, S. (2012). *Bürgerwille und Gebietsreform: Demokratieentwicklung und Neuordnung von Staat und Gesellschaft in Nordrhein-Westfalen 1965–2000.* München: Oldenbourg Verlag.

Nairn, T. (1997). *Faces of nationalism: Janus revisited.* London: Verso.

Paasi, A. (2012). Regional planning and the mobilization of "regional identity": from bounded spaces to relational complexity. *Regional Studies*, 47, 1206–1219.

Rausch, A.S. (2012). A framework for Japan's new municipal reality: assessing the Heisei gappei mergers. *Japan Forum*, 24, 185–204.

Rodríguez-Pose, A. & R. Sandall (2008). From identity to the economy: analysing the evolution of the decentralisation discourse. *Environment and Planning C*, 26, 54–72.

Smith, A. (1982). Ethnic identity and world order. *Millennium: Journal of International Studies*, 12, 149–161.

Spicer, Z. (2012). Post-amalgamation politics. *Canadian Journal of Urban Research*, 21, 90–111.

Studiegroep Openbaar Bestuur (2016). *Maak verschil: Krachtig inspelen op regionaal-economische opgaven.* Den Haag: Ministerie van Binnenlandse Zaken en Koninkrijksrelaties.

Terlouw, K. (2014). Lokale identiteit en schaalvergroting. *Openbaar Bestuur*, 24, 2–7.

Terlouw, K. & M. Hogenstijn (2015). *"Eerst waren we gewoon wij en nu is het wij en zij": gebruik slijtage en vernieuwing van regionale identiteiten.* Den Haag: Ministerie van Binnenlandse Zaken en Koninkrijksrelaties. www.rijksoverheid.nl/bestanden/documenten-en-publicaties/rapporten/2015/05/01/onderzoeksrapport-over-lokale-en-regionale-identiteiten/eerst-waren-we-gewoon-wij-terlouw-hogenstijn-2015.pdf.

Tomàs, M. (2012). Exploring the metropolitan trap. *International Journal of Urban and Regional Research*, 36, 554–567.

Van Assche, D. (2005). *Lokale politiek als katalysator van vertrouwen: binnengemeentelijke decentralisatie in Antwerpen*. Brugge: Vanden Broele.

Van Twist, M.J.W, M.S. Schulz, J. Ferket, J. Scherpenisse & M. van der Steen (2013). *Lichte evaluatie gemeentelijke herindeling: Inzichten op basis van 39 herindelingen in Gelderland, Limburg en Overijssel*. Den Haag: Ministerie van Binnenlandse Zaken en Koninkrijksrelaties.

Yack, B. (2012). *Nationalism and the moral psychology of community*. Chicago: University of Chicago Press.

2 Local identities conceptualised
From fixed facts to flexible discourses

Local identities are no longer regarded as fixed facts which can be objectively measured by skilled scientists. Local identities are now conceptualised as social constructions, which are constantly being reproduced and transformed by actors with different intentions. This chapter gives an overview of the main concepts, categorisations and models, which are used in subsequent chapters in the analysis of the usages of local identities during and after municipal amalgamation in two Dutch municipalities.

After discussing the traditional and now defunct conceptualisation of identities as fixed facts (§2.1), the current dominant conceptualisation of spatial identities as social constructions is discussed in §2.2. Then the relation between different types of communities and spatial identities is discussed in §2.3. This brings us to the ideal typical difference between "thick" and "thin" spatial identities (§2.4) and the gaps between how administrations and populations conceive local identities (§2.5). The role of spatial identities in the institutionalisation of spaces like municipalities and nations is discussed in §2.6. The layering of the identities of different spatial entities across scales when municipalities cooperate is examined in §2.7. The importance of spatial identities for the legitimation of amalgamated municipalities is the subject of §2.8. Resistance identities can emerge when the local population considers amalgamation as illegitimate. This chapter ends with a discussion on the changing relation between primary and secondary identities during municipal mergers.

2.1 Local identities as fixed facts with physical roots

Local and regional identities were traditionally seen as facts which were uncovered by the elaborate studies geographers made of a specific place or region. Until the 1960s, most geographers focussed on the specific way in which humans interacted with the environment within their region. Many traditional regional geographers followed the ideas of Alfred Hettner on how to study the interaction between different geographical layers. They started with physical layers, like soil and climate, and then moved upwards to human activities, like economy and culture. The painstaking study by the researcher of the locally very specific way in which all these material and human layers interact with each other would reveal its real identity. Identities of places and regions where thus formed through the long-term interaction between soil formation, cultivation, human settlements, and social, economic and territorial organisation. The different character of places, regions and nations where thus linked to very specific ways in which over time these interrelated layers were linked. The open or more hidden environmental determinism of this vision on local identity has, since the 1960s, been discredited in geography. Local identities are no longer regarded as fixed facts, but are conceptualised as social constructions (Holman 1995; de Pater *et al.* 2011; Johnston & Sidaway 2004).

This changing academic conceptualisation of what spatial identities are has only partially affected the popularity of environmental determinism in everyday narratives on local identities. The different identities of local communities are frequently linked by our interviewees to the different types of agriculture which used to dominate the daily life in the different local communities. One interviewee in Katwijk comments: "The characteristics of each village correspond to where they come from, what type of labour they did. That matches their specific temperaments."

The different local identities in the amalgamated municipality of Katwijk are commonly linked to the different means of existence in the different villages. Fishing at sea dominated in Katwijk aan Zee, horticulture in Katwijk aan de Rijn and the cultivation of flowers in Rijnsburg (see Figures 3.1 and 3.2).

The idea of historically formed fixed local identities is a popular argument used by many opponents of municipal amalgamations. The opposition against the amalgamation of the municipalities of the island of Goeree-Overflakkee in the old municipality of Goedereede is frequently linked to its history as a separate island. Its economy was traditionally dominated by fishers, while the other island, Overflakkee, was dominated by large-scale farming controlled by rich landowners employing poor farm workers. One interviewee remarks: "They are two islands, but they are really two separate islands, culturally."

This is despite the fact that they already became a single island in the eighteenth century. In 1751, a dam was constructed connecting the two islands and land between the islands was subsequently reclaimed.

2.2 The social construction of identities

The academic conceptualisation of local identities has however changed in the past decades. For instance, the leading Finnish geographer Anssi Paasi (2012, 3) stresses that a spatial identity is not a fixed fact, but a

social construct that is produced and reproduced in discourse. The discourses of regional identity are plural and contextual. They are generated through social practices and power relations both within regions and through the relationship between regions and the wider constituencies of which they are part.

Spatial identities are socially constructed representations of specific areas like places, regions or nations, and are produced and reproduced in discourse. A spatial identity is based on the communicated characteristics of that area in comparison to other areas and on the normative valuation of these differences. An area can have different identities based on the different representations and images of this area with different groups. The conceptualisation and appreciation of the identity of an area can especially vary between people living inside and outside an area.

To better understand what spatial identities are and how they are used, it is useful to first discuss the key characteristics of individual and collective identities. Individual identity gives meaning to the relation between the individual and the communities to which they belong. Identities are not fixed, but are fluid. During their life, an individual has to deal with new challenges in their relations with others. Individuals adapt their identity to make sense of this strained and changing relation between their individual uniqueness and their collective sameness. Individuals try to make sense of these conflicts through their construction of a more or less coherent life story (Verhaeghe 2014; Bauman 2004).

"identity" is revealed to us only as something to be invented rather than discovered; as a target of an effort, "an objective"; as something one still needs to build from scratch or to choose from alternative offers and then to struggle for and then to protect through yet more struggle.

(Bauman 2004, 15–16)

Collectives also have identities which likewise make sense of the strained and changing relation with yet larger entities. Spatial identities are not well-defined spatial facts, but are social constructs created and reproduced not through individual life stories, but through collective discourses shaped by the most powerful stakeholders. These spatial identity discourses become materialised in, for example, books, planning documents, newspaper reports or official websites (Paasi 2009; 2010; 2011; 2012). While these identity discourses are produced in spatial communities, we have to take a closer look at what spatial communities are and how these are linked to spatial identities. This is the subject of the next section.

2.3 Contingent communities and identities

Community is a much used but, according to Bernard Yack (2012), inadequately theorised concept. Social theory has traditionally focussed on closely knit village communities as the true and authentic type of community whose stranglehold on individual identity disappears in our modernising societies. But there are other kinds of communities. Communities are based on a shared characteristic which can vary from "a belief, a territory, a purpose, an activity, or merely the lack of a quality that some other group is thought to possess" (Yack 2012, 61).

Besides these cognitive aspects, communities are also based on affection, or what Yack calls social friendship and the conative aspects of support and solidarity.

There are, according to Yack, three basic types of communities based on three different shared characteristics. "The things that we share in different forms of community can be divided into three broad categories: the natural or necessary, the chosen, and the contingent" (Yack 2012, 78).

Natural communities are, for instance, families whose members are bound by close caring relations. Family members depend on each other for their livelihood and, especially in the case of babies, for their survival. Chosen communities are in contrast not based on nature, but on the deliberate choices of individuals to associate with likeminded individuals with whom they share similar values and objectives. These are voluntary associations of similar individuals who have chosen to belong to this community. People can also quite easily leave these chosen communities. Contingent communities on the other hand are based on involuntarily shared interests. Individuals are, for instance, born into territorial communities. Human existence is inextricably linked to the spaces people live in and which they have to share with others. This living together in space makes people dependent on each other for their quality of life (Blokland 2003, 78–79). Proximity, propinquity (Amin & Thrift 2002) or thrown-togetherness (Massey 2005) form the basis for spatial communities. In particular, individuals living together in the same political territory not only generate communal interests through their proximity, but they also develop their own ways to accommodate and regulate the different interests of partly unfamiliar and distant neighbours within their spatial community. These contingent communities lack the consistency and goal orientation of the other types of communities. Unlike natural communities, they cannot be based on causal necessity, and unlike chosen communities they are not based on conscious design (Yack 2012, 29).

> Communities that grow out of contingent forms of sharing are much harder to justify than those that owe their origins to nature or choice. For they lack both the sense of necessity of the former and the sense of rational purpose of the latter.
>
> (Yack 2012, 82)

Because of this ambiguous and conditional character of contingent communities, unlike natural and chosen communities, they do not have a clear identity which justifies them to their members and which sets them clearly apart from other contingent spatial communities. They arduously have to construct their identity by selectively combining elements on which the identity of natural and chosen communities are based. They combine descent and heritage from the former, with choice and consent from the latter (Yack 2012, 45, 59). They combine the focus on the past, which is more present in natural communities, with the focus on the future, which is a characteristic more of chosen communities.

2.4 The ideal typical contrast between thick and thin spatial identities

Making a distinction between thick and thin spatial identities can further clarify the hybrid character of the identity of contingent spatial communities. Thick spatial identities are more backward-looking and value the spatial community as a political goal in itself. They focus more on bonding within a community, while thin identities focus more on bridging between communities. Thin spatial identities are more forward-looking and value more the effectiveness of their especially economic policies. Moreover, thin spatial identities are more functional and linked to sectorial policies and special interests and stakeholders, while thick spatial identities are more integrative. Thin spatial identities are thus created around a few – often economic – characteristics, while thick spatial identities cover a broad range of cultural, social, political, environmental and economic characteristics. Thin spatial identities are more changeable. Their spatial form and meaning can be adapted to changing circumstances. They are less based on static territories with a fixed meaning, but focus more on fluid networks and dialogue (Terlouw 2009; Delanty & Rumford 2005,

68–86; Bauman 2004, 13–46; Antonsich 2011; Sack 1997; Jones & MacLeod 2004). Table 2.1 gives an ideal typical overview of the differences between thick and thin spatial identities. In reality, identity discourses will always mix thick and thin identity elements into a more or less coherent spatial identity discourse. Even actors in regions with a very thick and well-established identity, like Catalonia and Scotland, will always try to link their thick identity rooted in history with thin identity elements. Their regional identity discourses link the past with the future. Conversely, stakeholders involved with newly created regions will also attempt to link up with elements of more established thick spatial identities. Newer forms of regional cooperation, such as metropolitan regions, try to thicken their thin economic regional identity discourse by referring to a glorious past, even though this was long before the administrations in these regions started to cooperate. They do this to widen their support base from policy makers to the general population (Terlouw 2012). Regional administrations sometimes use different regional identities for different audiences. Thin regional identities focussing on economic competitiveness are more used to attract outside investors. Thick regional identity discourses are frequently used as an ideological shield to conceal the drawbacks of these neoliberal policies for the general population, by focussing on the supposedly shared interests of all members of a territorial community (Cox 1999).

These different types of identity discourses are however not fixed. Not only can the particular thick and thin identity elements used change, but also the area to which they refer is subject to political debate and can change over

Table 2.1 The difference between thick and thin spatial identities

Aspect	Ranging from thick:	to thin:
Spatial form	Closed Territorial	Open Network
Organisation Participants	Institutionalised General population	Project Administrators and specific stakeholders
Purpose	Broad and many Culture	Single Economy
Time	Defensive Historical oriented Stable Old	Offensive Future oriented Change New
Scale focus	Local and national	Global

Source: Terlouw (2009).

time. Inhabitants especially tend to value more the thick identity elements with which they are familiar, while the administration tends to focus more on new economic developments linked to other thinner identity elements in wider networks. This is further discussed in the next section.

Box 2.1 Emsland: combining thick and thin elements in a new regional identity discourse

The Emsland is a region in the northwestern corner of Germany, on the border with the Netherlands. Its administration presents a regional identity based on a balanced mixture of thick and thin elements. The regional identity of the Emsland has changed over the last half century from a very thick and quite negative regional identity of a peripheral region in Germany, to a positive regional identity which mixes traditional thick rural elements with new thin elements based on its economic success and competitive position in Europe.

The communicated identity of the Emsland was until the 1950s dominated by its remoteness and barren landscape. Its inaccessible parts like the Hümmling were for centuries a refuge for outlaws and Gypsies (Knottnerus 1992, 37; Nauhaus 1984). Combined with images of its unproductive landscape, riddled with bogs, heaths and sand drifts and almost without villages, this engraved a negative image of the Emsland in the local and German collective consciousness. The Emsland was the iconic German periphery, waiting to be developed through modernisation (Hucker *et al.* 1997, 348; Knottnerus 1992; Niehoff 1995).

The negative character of this thick regional identity discourse stimulated the central and regional authorities to modernise the Emsland. Marshes were cleared, new agriculture land was cultivated, its infrastructure was improved and many industries were attracted with subsidies. The Emsland lost a lot of its specific characteristics and became more like the rest of rural Germany (Nauhaus 1984; Niehoff 1995; Schüpp 1992).

For the new administrative district of Emsland, created in 1977, this economic development became an important part of its regional identity discourse. They stress that the Emsland is no longer "the poorhouse of the nation", but that the Emsland had over the last decade one of the highest regional economic growth rates in Germany. It is no longer an isolated peripheral region at the German–Dutch border, but an accessible and competitive region in the heart of Europe (Niehoff 1995).

The regional identity discourse of the administration of the Emsland combines thin and thick identity elements. It boasts its economic performance rooted in the regional networks of its internationally competitive companies. But it also uses thick elements like its landscape in new ways. The old unproductive wastelands are now re-valued as an idyllic rural landscape with ample room for recreation and suburban housing (Terlouw 2012).

2.5 Identity gaps between administrations and populations

There are frequently gaps between the visions of the population and the administration on what constitutes the identity of their place, municipality or region. The population especially tends to value more the thick identity elements with which they are familiar, while the administration tends to focus on new economic developments. The nature of these identity gaps can be further clarified using a more detailed classification of spatial identities. Table 2.2 differentiates between *communicated, conceived, ideal* and *desired* identities. This distinction is based on a method developed by John Balmer and Stephen Greyser (2002) to communicate company brands more effectively through bringing these different identities in line in the formulation of a coherent identity discourse. Their approach has also been applied to the branding of cities, regions and countries (Trueman *et al.* 2004; Kavaratzis & Ashworth 2005; Kaplan *et al.* 2010). Table 2.2 depicts the dominant relations between these types of identity, their time perspective and their relation with different stakeholders.

The *communicated* identity is based on a representation of established characteristics. In particular, the official communications of administrations selectively use those qualities within their area which fit their political aims. Communicated identities are presented through a wide range of messages. This covers not only the explicit marketing and branding campaigns to attract visitors, but also the identity implicitly used in, for instance, policy documents. The effectiveness of this communicated identity is based on the correspondence with the *conceived* identity, which is based on the perception of the region by different groups within and outside the area. The *ideal* identity is the goal of the policies of the administration. The ideal identity either amends current deficits or adapts the region to the expected changes in its environment, which are at this time frequently linked to the pressures generated by increased global competitiveness. This thin ideal identity promoted by administrations has to link up with the thicker *desired* identity of the population in order to legitimise

Table 2.2 Identity types, stakeholders and time perspective

Stakeholders	Time perspective	
	Perceived (Backward)	*Aspired (Forward)*
Administration	Communicated	Ideal
Population	Conceived	Desired

Source: adapted from Terlouw (2014).

these new policies. Whereas the ideal identity focusses on the efficient realisation of specific policy goals in relation to the outside world, the desired identity is thicker and much broader. The desired identity is based on the vision of a better future for the whole population and rooted in their established values and norms.

A spatial community functions most easily when these different types of identity are aligned. Gaps between the different types of identity referring to the same space hinder collective action in a spatial community, both internally and externally. In particular, differences in the aspired identities of the administration (ideal identity) and the population (desired identity) concerning the wanted future of their community frequently fuel political controversies. These political differences on future development can also widen the gap between the communicated and conceived identities through the reinterpretation of their perceived identity by the population and can thus change how identities are communicated. (How this can result in the formation of resistance identities is discussed in §2.8.) Not only do these identity gaps between population and administration within a community reduce the ability to develop strong policies, but they also restrict the ability of a community to successfully promote its interest towards other communities. It makes communities vulnerable to both fragmentation and amalgamation.

There are several strategies to bridge these identity gaps. Balmer and Greyser (2002) stress the importance to start with detecting the differences between the existing different types of identity. They then devise a strategy to adapt these different identities to create a coherent brand. Although their main instrument to achieve this is a communication strategy – this is the fastest and easiest way to align identities – they also acknowledge that this sometimes has to be supplemented by material changes. In city marketing and branding, the importance of physical changes for the promotion of a city is also increasingly emphasised. Communicating an improved image of a city is only effective when these are linked to physical improvements. According to Mihalis Kavaratzis (2004), this even constitutes the primary form to communicate the identity of a city. The story of a better place or space is only effective when the communication of this image is linked to recognisable improvements in the material fabric of that place or space. This is a more convincing form of communication than only using media, but takes more time and effort (Anholt 2010).

2.6 Spatial identities and the institutionalisation of spaces

Spatial identities are not fixed in time, but are formed over time. They develop in relation to other ways in which humans organise spaces. Spatial identities are part of the wider process in which spaces become institution-alised. Anssi Paasi (1986, 1991, 1996, 2012) uses four distinct but inter-related aspects or "shapes" to analyse this process of the institutionalisation of spaces. Through the combination of its territorial, symbolic, institutional and functional shape, an area becomes institutionalised in its own specific way.

The territorial shape is the most tangible aspect. It includes the borders and the way in which these were constructed in history. Physical spatial characteristics like landscape and land use patterns are also part of this ter-ritorial shape. The spatial stereotypes partially based on this territorial shape and on the characteristics of its population are a common source of the symbolic shape of an area. This forms its spatial identity. The institu-tionalisation of this symbolic shape of an area is also organised. Adminis-trations and civil society constantly communicate this spatial identity through, for instance, educational institutions, policy documents and the mass media. This forms together with the political administration of its ter-ritory the institutional shape of an area. The functional shape refers to the established role of an area in larger systems. These are based on, for instance, its economic ties with neighbouring areas, or its place in the administrative national hierarchy. When these four shapes interlock, they reinforce each other and generate institutionalised areas – like municipal-ities, regions or nations – with a strong spatial identity.

Box 2.2 Stuck in a thick regional identity: Lippe

Lippe, a region in the northeast of the German federal state of Nordrhein-Westfalen, has a very strong traditional regional identity. Historic buildings, a wooded hilly landscape and the statue of Hermann – the German victor over the Roman legions – dominate the communicated identity of the administrative region of Lippe.

Lippe is a strongly institutionalised region based on a territorial shape formed in the late Middle Ages and which has hardly changed over the cen-turies. Its rural landscape is hardly touched by industrialisation and urbani-sation. Its institutional shape is also strong. After centuries of being an independent state, it is now an administrative district. Its symbolic shape has only become stronger over the years and has focussed more and more on a thick regional identity. Its functional shape is more problematic. Unlike its Prussian neighbouring regions, it did not industrialise in the nineteenth and twentieth centuries. After the Second World War, Lippe's economy profited from the "white industry" of the health spas funded by the German welfare state. However, in the past decades, it has suffered from the reductions in welfare spending by the German state, which decimated the funding of visits to health spas. Not only its functional shape, but also its institutional shape is now threatened by administrative reforms benefitting other regions. Its thick regional identity discourse focussing on its history and the landscape within its territory hinder Lippe in effectively participating in new forms of regional cooperation, like Ostwestfalen-Lippe (OWL), discussed in Box 2.3, which has developed a thinner regional identity discourse which is more open, networked, future oriented and focussed on economic competitiveness (Terlouw 2012).

The identities of institutionalised areas are based on stable communities with collective identities which are passed on from generation to generation. The recent scaling up of social and economic relations undermines this. The position of different areas in the international division of labour becomes more changeable. Globalisation especially changes functional shapes rapidly and thus undermines the institutionalisation of areas. Transformations of the state organisation, through, for instance, the emergence of new forms of administrative cooperation, or the creation of new regions through amalgamation, further undermine institutionalised spaces. Through the increased pace of these changes, there is less and less time for the institutionalisation of areas and for thick spatial identities to take root.

Box 2.3 Creating a thin neoliberal economic regional identity: OWL

In Bielefeld, a large city in the northeast of the German federal state Nordrhein-Westfalen and bordering the district of Lippe, important regional companies together with the regional chambers of commerce of Bielefeld and Lippe founded, in 1989, an association to promote the region Ostwestfalen-Lippe (OWL) as a business location. As its complicated name suggests, OWL has no established regional identity. OWL wants to communicate a regional identity which is different from both the more rural Westfalen and the declining industrial Ruhr area. By stressing its many mid-sized cities with thriving local companies, it communicates a distinct thin regional identity. Its regional identity discourse is thus based on economic characteristics, and is offensive and future oriented. It focusses its communication not on its own population, but on entrepreneurs. The communicated regional identity has shifted over time from correcting a negative image to promoting OWL as an innovative and cooperative business community showing the rest of Germany how to improve global competitiveness through networking and deregulation (Terlouw 2012).

The outcome of the institutionalisation of spaces like municipalities, administrative regions and nation-states has in recent times become more variable and unstable. The institutionalisation processes of different spaces are increasingly linked and intertwined. Our world is becoming less a neat mosaic of separate spaces, but increasingly a hotchpotch of overlaying, overlapping and intermingled spaces. New spaces constantly emerge, challenging and sometimes replacing older spaces. The institutionalisation of new spaces can result in the co-institutionalisation or de-institutionalisation of older spaces.

Co-institutionalisation means that institutionalised spaces are transformed through their linkages with these new spaces. Their established thick identities are, for instance, reinterpreted and linked to others, through the development of an overarching thinner spatial identity discourse using also a selection of shared thicker identity elements. The next section, §2.7, discusses how these layered identities of new but overlapping spaces are formed through the selective uploading and downloading from identity elements of more established smaller and larger spaces (Terlouw & van Gorp 2014). This section shows how co-institutionalisation can bridge long-established thick spatial identities through their overlay with thinner identity discourses. In the subsequent section, §2.8, the focus shifts to the role of identities in the de-institutionalisation of established spaces and discusses how this can result in the formation of resistance identities.

2.7 The layering of spatial identities

Newly formed regions and municipalities have to relate their identity discourses to those of many other spaces in the increasingly complex administrative environment. The spatial identity discourses of these new spaces have to link up with elements of the thicker identities of more established spaces. Identity elements can be selected from smaller localities, from (partially) overlapping areas, or from larger regions or states. This results in complex and changing relations between these spatial identity discourses, which forms more a kind of "breccia" (Bartolini 2011) – a coarse grained sedimentary rock made of stone fragments and cemented together by finer material – than a layer cake with the nation as the icing on top. However, to analyse how spatial identities are linked, we use the metaphor of layered scales as depicted in Table 2.3. This model of layered spatial identity discourses is a Weberian ideal type, which is helpful to better understand the much more complicated relations between spatial identities. Simple and clear abstractions can help us to better understand the complex and messy reality. Table 2.3 is based on a simplification of the relation between the properties of different levels of analysis (Hoekveld &

Hoekveld-Meijer 1994; Lazarsfeld & Menzel 1961). The characteristics of regions are not only based on characteristics specific to the regional level, but also constructed from its constituent elements (like places) and its belonging to larger collectives (like large regions or the nation-state). These spaces are not as neatly hierarchically nested as Table 2.3 suggests. In reality they frequently overlap and intermingle. Co-institutionalisation of new regions in relation to more established areas involves the layering of spatial identities. This is done through the creation of spatial identity discourses in which the newly created regions or municipalities selectively use some of the characteristics of other, related spatial identities of more established institutionalised spaces.

Table 2.3 distinguishes between negative opposition and positive associations. The importance of negative oppositions or "othering" for the construction of spatial identities has received a lot of attention in the past few decades in human geography (Cloke & Little 1997; Sibley 1995; Wilton 1998). However, positive associations or "brothering" are also important in regional identity discourses, especially for those new regions which try to mobilise support for new policies and want to justify their policies (Boisen *et al.* 2011; Calhoun 1994; Donaldson 2006; Zimmerbauer 2011). These positive and negative connotations not only focus on neighbouring regions, but can also cross scales. The resultant layering in the identity discourse of a new space can be based on some characteristics of its elements (uploading), or characteristics of the larger collective to which a new space belongs (downloading). For instance, the innovative identity of high-tech regions is strengthened by using images of science parks (uploading) and by using national characteristics, like the Dutch

Table 2.3 The layering of spatial identities

Scale	Positive association	Negative oppositions
Collective	Downloading of positive general characteristics	Antagonism towards negative general characteristics and centralisation
Neighbours	Cooperation and aligning with similar and strong regions	*New space* (e.g. amalgamated municipality)	Opposition towards "other" regions
Elements	Uploading of positive local elements	Exclusion of unwanted elements

Source: adapted from Terlouw & van Gorp (2014).

mercantile spirit or German craftsmanship (downloading). This selective layering of identity can skip scales when the identities at those scales are not in line with the desired identity of the new regions. Many communications of peripheral border regions focus therefore on their "Europeanness" rather than on their marginal position in their nation-state. This layering in the identity discourses is very important for the legitimation of these new spaces and their policies. The established legitimacy of the national government and municipalities with their elected councils is tapped into by downloading and uploading some of their characteristics in the identity discourse of new spaces.

Box 2.4 The layered identity discourse of BrabantStad

In BrabantStad the five large cities of Noord-Brabant and the provincial administration have cooperated since 2001 to stimulate economic development. These cities in the south of the Netherlands are regarded as crucial for the economic development of the whole province. BrabantStad, for instance, initiates projects in which local companies cooperate, it collectively promotes the cities to international companies and, last but not least, this cooperation is used to attract more national funding for their economic development.

The network of BrabantStad is not an established region, but a new space made up of a functional network very similar to a functional overlapping competing jurisdiction (FOCJ), which will be discussed in Box 6.2 (page 100). As such, it has only a very thin economic identity, which it tries to connect to more established spatial identities. For instance, in its communications with the outside world, BrabantStad downloads from the national level the language skills of the Dutch and it uploads the aura of the high-tech campus in Eindhoven linked to the facilities of the Philips company. It also aligns itself with the neighbouring Green Forest region, which was recently created to protect and develop the recreational and residentially attractive green area between the five cities of BrabantStad. BrabantStad's identity discourse also links up with the traditional antagonism between the southern peripheral province of Noord-Brabant and the urbanised core region of Holland in the western part of the Netherlands. It however excludes the deprived working class neighbourhoods in their cities. BrabantStad thus strengthens its very thin economic identity by aligning itself with more established spatial identities at different layers (Terlouw & van Gorp 2014; van Gorp & Terlouw 2016).

2.8 Identity and legitimacy: de-institutionalisation and the formation of resistance identities

Political spaces are not only formed but can also disappear. Established spaces like municipalities can de-institutionalise when newer spaces take over some or all of their powers. This can be the case when municipalities cooperate intensively or when municipalities are amalgamated. This can generate popular resistance while

> communities of free and equal citizens cannot live so easily with continual adjustments of their boundaries. For these communities need to know exactly whom the administration is serving and where the limits of its authority lie. Without sharp and relatively settled boundaries the liberal project of rendering coercive authority accountable to the people who created it is impossible.
>
> (Yack 2012, 32)

A reorganisation like an amalgamation can create, in a relatively short period, new territorial borders, new political institutions and a new functional shape. However, it takes much more time for a symbolic shape of this new territory to become institutionalised. This lack of an established spatial identity is a weak point which is frequently used by opponents who try to hang on to their old familiar spaces.

Changes in the administrative territories and their hierarchical relations challenge the legitimation of power. The coexistence of multiple new regions in the same area hinders them acquiring legitimacy among their population. The lack of proper popular representation especially is seen as a democratic deficit which undermines the legitimacy of the administrative cooperation in new regions (Boudreau 2007; Brenner 2004; de Vries & Evers 2008; Teisman 2007). These concerns are based on Max Weber's views on the legitimation of power based on popular acceptance. David Beetham, a leading political philosopher and political theorist on the legitimation of power, argues however that legitimacy involves more than the belief of the population in the legitimacy of a political system. Legitimacy is not based on public opinion, but is the result of the correspondence between the political system and social norms (Beetham 1991, 8).

According to Beetham, legitimacy is based on the coherent but changeable combination of three dimensions: legality, expressed consent and justifiability. Legality refers to adherence to the established rules of acquiring and exercising power. The expressed consent of the population with the power structures in society is either mobilised, through, for instance, oaths and the participation in mass events, or it results from elections. Justifiability is based on social norms on the source of political authority and the purpose of government. State power "must derive from a source that is acknowledged as authoritative within society; it must serve ends that are recognised as socially necessary, and interests that are general" (Beetham 1991, 149).

Justifiability is based not only on the source, but also on how successful administrations are in serving a communal interest. Legitimate political systems must adequately and efficiently serve these socially defined common interests (Beetham 1991, 70, 86). This common interest is linked to the values and identity of that community.

> (T)he legitimation of power rules is not only the development and dis-semination of an appropriate body of ideas, or ideology, but the con-struction of a social identity by a complex set of often unconscious processes, which make that identity seem "natural", and give the justifying ideas their plausibility.
>
> (Beetham 1991, 78)

Beetham's approach helps us to better understand the importance of local identities in the legitimation of power at the local level. Even during the heydays of the nation-state, municipalities were important for the regulation and policing of social life, the representation of local interests and the provision of public services (Flint & Taylor 2007). The legitimacy of smaller municipalities especially is now being undermined by the growing number of tasks they are not equipped to perform. The current rescaling of western states combines therefore the decentralisation of responsibilities to the local level, with the centralisation from the local to the regional level through their cooperation in new regions or municipal amalgamations (Brenner 2004). Increased efficiency and effectiveness justifies both forms of administrative restructuring, but their legitimacy is undermined on other dimensions of legitimation. Through municipal restructuring, established political systems are profoundly changed, which undermines their legality as they do not yet have established rules of acquiring and exercising power. It is also difficult to legitimise municipal restructuring through the expressions of popular consent. Although the population in the restructured municipality will elect a council, municipal restructuring is often contested by large sections of the affected population.

Amalgamation is frequently seen as an external threat to well-established local identities. This can lead to the development of a resistance identity discourse bonding local inhabitants by focussing on the old municipal territory, its historic roots and its difference from others (Castells 2010; Zimmerbauer *et al.* 2012; Zimmerbauer & Paasi 2013). The focus of local resistance identity discourses shifts from the outside to the inside and from the future to the past. The diminishing institutional and functional shape can thus go hand in hand with a strengthening of its symbolic and territorial shape. Paradoxically, the demise of territorial autonomy can coincide with the growth of the symbolic importance of these borders. The process of spatial institutionalisation is reversed in a way. Existing narratives on the history and culture of the de-institutionalising space are reinterpreted and refocussed on the historical meaning of its borders. The differences with neighbouring communities are further emphasised and get stronger emotional overtones. Forced de-institutionalisation can thus result in the thickening of existing spatial identities discourses and their transformation into resistance identities. Box 2.5 discusses in more detail the role of a local resistance identity in the amalgamation of the Finnish municipality of Nurmo.

Box 2.5 Resistance against municipal amalgamation in Nurmo, Finland

In Finland the national government is also concerned about the efficiency and effectiveness of small municipalities. Over the last decade, the number of municipalities in Finland has been reduced from 436 to 342. These amalgamations frequently met with strong opposition from the local population. For instance, in the western part of Finland, the vast majority of the population of Nurmo opposed the merger with the provincial capital Seinäjoki even though many of its inhabitants work there.

The administration of Seinäjoki took the initiative to merge with the suburban municipality Nurmo and the rural municipality Ylistaro. In 1992, an earlier attempt at amalgamation failed through the opposition from Nurmo. To oppose the new amalgamation attempt in 2006, concerned inhabitants founded the action group ProNurmo. The majority of its members were citizens who were not previously involved in local politics. They organised meetings and communicated that this merger was not inevitable through their website, publications and local media. They successfully persuaded the local political parties to organise a referendum. Despite 63.1 per cent voting for independence, Nurmo's municipal council voted for amalgamation with a one-vote majority.

Local identity was a key argument used by the opponents of amalgamation. ProNurmo activists stressed that Nurmo's local identity was very different from Seinäjoki. But when interviewed, they had difficulty explaining what characterised their local identity. Their quite general and vague references to local identities were part of their general objections to the scaling-up of local administrations and specifically their resistance against merging with Seinäjoki. They mirrored the good characteristics of their rural local identity with the bad characteristics of a city. They formulated a resistance identity discourse based on a nostalgic image of a tight local community that had developed its own way of life and shared a common history. They contrasted this with the feared urban future, which would make them anonymous and in which their special interests and preferences would be neglected by the large urban administration. They feared a loss of autonomy and control over their daily life. On the other hand, the supporters of amalgamation denounced the opponents as living in the past and appealing to emotions instead of using rational arguments (Zimmerbauer *et al.* 2012: Zimmerbauer & Paasi 2013).

2.9 Identities and scale: primary and secondary identities

Before amalgamation, local communities were familiar with how the municipal administration addresses their interests and handles their identities. Amalgamations undermine these certainties. This is a problem all organisations face during mergers. Uncertainty and the feared loss of the identity of one's old organisation can generate resistance to a merger, resulting in a lack of identification and involvement with the merged organisation (Burchhardt 2015; van Knippenberg *et al.* 2002). Uncertainty about the future of the new organisation can even strengthen the identification with the pre-merger organisation (Giessner 2011).

The formulation of a transitional identity built on the legacy of the old organisational identities, but linked to the goals of the new organisation, can give a sense of continuity and thus reduce the uncertainties of the merger process. This can be the basis for the formation of a new identity for the new organisation (Drori *et al.* 2013; van Knippenberg *et al.* 2002; Roundy 2010). Clear procedures are also very important to reduce these uncertainties (Gleibs *et al.* 2008).

Dealing with other identities is always part of an identity discourse. Local identity discourses are always related to other competing or complementary discourses. Local identities are not fixed facts which exist in isolation, but emerge out of the confrontation with other identity discourses. According to Niklas Luhmann (1983), the procedures to deal with such differences create legitimacy in a social system. Through interaction and communication between different stakeholders in a social system – like the political system in a municipality – an "open" identity is formed based on the locally developed restrictions on political decision making (Luhmann 1983, 42). Legitimation in a social system is not based on similarity and consensus, but on procedures through which different established interests and identities are given due consideration in the political decision making process (Luhmann 1983, 196–197). The established procedures of dealing with these differences also create a sort of identity (Luhmann 1983, 199). When social systems become more complex – through, for instance, municipal amalgamations – it becomes more difficult for such a secondary identity to emerge. It becomes more difficult for procedures to become established to deal legitimately with all different interests and identities in the decision making process (Luhmann 1983, 234). Thus two different types of identities are important during amalgamations. First of all, there are the distinct and established local identities. These are the primary local identities which are relatively stable and well known by the local population. Second, there are the different ways in which communities deal with these primary identities. Over time, communities learn and institutionalise ways to deal with different local identities. They develop informal procedures to deal with the shared and diverging interests and identities in their community. This forms a kind of secondary identity, which on the one hand focusses on the common goals and is linked to the shared elements in their different local identities. On the other hand, their secondary identity is based on the ways in which they have learned over time to avoid confrontations over issues which affect the most sensitive elements in their different collective identities. The specific way of dealing with these different identities becomes a kind of secondary overarching local identity in a municipality.

Local identity discourses focus not only on the primary local identities, but also on the secondary identity of how to deal with these primary identities in daily life and in municipal politics. While primary identities are more based on distinctiveness and differences, secondary identities are based on bridging these differences. Changes in the power balance between local groups and external influences make secondary identities less stable than primary local identities based on community characteristics. Figure 2.1 depicts some of these relations.

This relation between primary and secondary identities can also explain why sometimes local resistance identities emerge during municipal amalgamations, as discussed in §2.8. During an amalgamation, this secondary identity disappears with the old municipality. The disappearance of the protective shield of a secondary identity exposes the underlying primary local identities, and can bring local identities into the centre of the local political debate. Amalgamation is then perceived as an external threat which undermines local identity. This can lead to the development of a resistance identity discourse which bonds its inhabitants by focussing on the old territory, its historic roots and its difference from others. The focus in such local identity discourses thus shifts from the outside to the inside and from the future to the past.

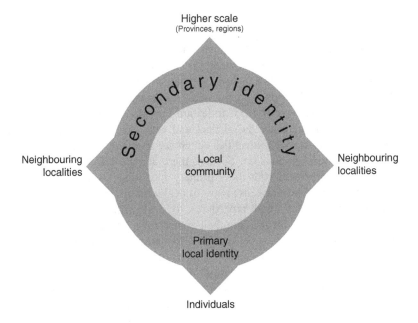

Figure 2.1 Local identity and scale.

This distinction between primary and secondary identities can also be linked to the layering of spatial identities by new regions, as discussed in §2.7. These new region sometimes emerge when the secondary identity of a municipality is too weak and indistinct to support the primary local identities. Regional cooperation or municipal amalgamations can then help to promote a new, more attractive, secondary, thinner, more regional identity, based on a selection of characteristics of established primary local identities.

References

Amin, A. & N. Thrift (2002). *Cities: reimagining the urban.* Cambridge: Polity.

Anholt, S. (2010). *Places: identity, image and reputation.* London: Palgrave.

Antonsich, M. (2011). Grounding theories of place and globalization. *TESG*, 102, 331–345.

Balmer, J.M.T. & S.A. Greyser (2002). Managing the multiple identities of the corporation. *Californian Management Review*, 44, 72–86.

Bartolini, N.L.M. (2011). *Modernizing the ancient: brecciation, materiality and memory in Rome.* Milton Keynes: Open University.

Bauman, Z. (2004). *Identity: conversations with Benedetto Vecchi*. Cambridge: Polity.

Beetham, D. (1991). *The legitimation of power*. Basingstoke: Macmillan.

Blokland, T. (2003). *Urban bonds*. Cambridge: Polity Press.

Boisen, M., K. Terlouw & B. van Gorp (2011). The selective nature of place branding and the layering of spatial identities. *Journal of Place Management and Development*, 4, 135–147.

Boudreau, J.-A. (2007). Making new regions: mobilizing spatial imaginaries, instrumentalizing spatial practices, and strategically using spatial tools. *Environment and Planning A*, 39, 2593–2611.

Brenner, N. (2004). *New state spaces: urban governance and the rescaling of statehood*. Oxford: Oxford University Press.

Burchhardt, S. (2015). *Competition with identity driven entry: a principal multiagent model on the success of mergers and acquisitions*. Wiesbaden: Springer.

Calhoun, C. (1994). Social theory and the politics of identity. In: C. Calhoun (Ed.) *Social theory and the politics of identity*. Cambridge: Blackwell.

Castells, M. (2010). *The power of identity* (second edition). Chichester: Wiley-Blackwell,

Cloke, P. & J. Little (Eds) (1997). *Contested countryside cultures: otherness, marginalization and rurality*. London: Routledge.

Cox, K.R. (1999). Ideology and the growth coalition. In: J.A. Wilson (Ed.) *The urban growth machine: critical perspectives, two decades later*. New York: SUNY.

De Pater, B., O. Atzema, R. Boschma, P. Druijven, P. Groote, B. van Hoven, V. Mamadouh & K. Terlouw (2011). *Denken over regio's: geografische perspectieven*. Bussum: Coutinho.

De Vries, J. & Evers, D. (2008). *Bestuur en ruimte: de Randstad in internationaal perspectief*. Den Haag: Ruimtelijk Planbureau.

Delanty, G. & C. Rumford (2005). *Rethinking Europe: social theory and the implications of Europeanization*. London: Routledge.

Donaldson, A. (2006). Performing regions: territorial development and cultural politics in a Europe of the Regions. *Environment and Planning A*, 38, 2075–2092.

Drori, I., A. Wrzesniewski & S. Ellis (2013). One out of many? Boundary negotiation and identity formation in postmerger integration. *Organization Science*, 24, 1717–1741.

Flint, C. & P. Taylor (2007). *Political geography: world-economy, nation-state and locality*. Harlow: Pearson.

Giessner, S.R. (2011). Is the merger necessary? The interactive effect of perceived necessity and sense of continuity on post-merger identification. *Human Relations*, 64, 1079–1098.

Gleibs, I.H., A. Mummendey & P. Noack (2008). Predictors of change in postmerger identification during a merger process: a longitudinal study. *Journal of Personality and Social Psychology*, 95, 1095–1112.

Hoekveld, G.A. & G. Hoekveld-Meijer (1994). Regional development in spatial and social contexts: key concepts of a regional geographical methodology. In:

K. Terlouw (Ed.) *Methodological exercises in regional geography: France as an example.* Utrecht: KNAG.

Holman, H. (1995). What's new and what's regional in the new regional geography. *Geografiska Annaler B*, 77, 47–63.

Hucker, B.U., E. Schubert & B. Weisbrod (Eds) (1997). *Niedersächsische Geschichte.* Göttingen: Wallstein Verlag.

Johnston, R.J. & J.D. Sidaway (2004). *Geography and geographers: Anglo-American human geography since 1945.* London: Routledge.

Jones, M. & G. MacLeod (2004). Regional spaces, spaces of regionalism: territory, insurgent politics and the English question. *Transactions IBG*, 29, 433–452.

Kaplan, M.D., O. Yurt, B. Guneri & K. Kurtulus (2010). Branding places: applying brand personality concept to cities, *European Journal of Marketing*, 44, 1286–1304.

Kavaratzis, M. (2004). From city marketing to city branding: Towards a theoretical framework for developing city brands. *Place Branding*, 1, 58–73.

Kavartzis, M. & G.J. Ashworth (2005). City branding: an effective assertion of identity or a transitory marketing trick? *TESG*, 96, 506–514.

Knottnerus, O.S. (1992). Räume und Raumbeziehungen im Ems Dollart Gebiet. In: O.S. Knottnerus, P. Brood, W. Deeters & H. van Lengen (Eds) *Rondom Eems en Dollard.* Groningen/Leer: Schuster Verlag.

Lazarsfeld, P.G. & H. Menzel (1961). On the relation between individual and collective properties. In: A. Etzioni (Ed.) *Complex organizations: a sociological reader.* New York: Holt.

Luhmann, N. (1983). *Legitimation durch Verfahren.* Frankfurt am Main: Suhrkamp.

Massey, D. (2005). *For space.* London: Sage.

Nauhaus, K.-E. (1984). *Das Emsland im Ablauf der Geschichte.* Sögel: Emsländische Landschaft.

Niehoff, L. (1995). Vom "Armenhaus Deutschlands" zum Landkreis mit hoher lebensqualität. In G. Müller (Ed.) *Emsland.* Oldenburg: Verlag Kommunikation und Wirtschaft.

Paasi, A. (1986). The institutionalisation of regions: a theoretical framework for understanding the emergence of regions and the constitution of regional identity. *Fennia*, 164, 105–146.

Paasi, A. (1991). Deconstructing regions: notes on the scales of spatial life. *Environment and Planning A*, 23, 239–254.

Paasi, A. (1996). *Territories, boundaries and consciousness: the changing geographies of the Finnish-Russian border.* Chichester: Wiley.

Paasi, A. (2009). The resurgence of the "region" and "regional identity": theoretical perspectives and empirical observations on regional dynamics in Europe. *Review of International Studies*, 35, 121–146.

Paasi, A. (2010). Regions are social constructs, but who or what "constructs" them? Agency in question. *Environment and Planning A*, 42, 2296–2301.

Paasi, A. (2011). The region, identity, and power. *Procedia Social and Behavioral Sciences*, 14, 9–16.

Paasi, A. (2012). Regional planning and the mobilization of "regional identity": from bounded spaces to relational complexity. *Regional Studies*, 47, 1206–1219.

Roundy, P.T. (2010). Gaining legitimacy by telling stories: the power of narratives in legitimizing mergers and acquisitions. *Journal of Organisational Culture, Communications and Conflict*, 14, 89–105.

Sack, R. (1997). *Homo geographicus*. Baltimore: John Hopkins University Press.

Schüpp, H. (1992). Zur Bedeutung der Emsland GmbH für die wirtschaftliche Entwicklung des Emslandes nach 1945 In: O.S. Knottnerus, P. Brood, W. Deeters & H. van Lengen (Eds) *Rondom Eems en Dollard*. Groningen/Leer: Schuster Verlag.

Sibley, D. (1995). *Geographies of exclusion*. London: Routledge.

Teisman, G. (2007). *Stedelijke netwerken: ruimtelijke ontwikkeling door het verbinden van bestuurslagen*. Den Haag: NIROV.

Terlouw, K. (2009). Rescaling regional identities: communicating thick and thin regional identities. *Studies in Ethnicity and Nationalism*, 9, 452–464.

Terlouw, K. (2012). From thick to thin regional identities? *Geojournal*, 77, 707–721.

Terlouw, K. (2014). Iconic site development and legitimating policies: the changing role of water in Dutch identity discourses. *Geoforum*, 57, 30–39.

Terlouw, K. & B. van Gorp (2014). Layering spatial identities: the identity discourses of new regions. *Environment and Planning A*, 46, 852–866.

Trueman, M., M. Klemm & A. Giroud (2004). Can a city communicate? Bradford as a corporate brand. *Corporate Communications*, 9, 317–330.

Van Gorp, B. & K. Terlouw (2016). Making news: newspapers and the institutionalisation of new regions. *TESG*, DOI:10.1111/tesg.12209 (forthcoming).

Van Knippenberg, D., B. van Knippenberg, L. Monden & F. de Lima (2002). Organizational identification after a merger: a social identity perspective. *British Journal of Social Psychology*, 41, 233–252.

Verhaeghe, P. (2014). *What about me? The struggle for identity in a market-based society*. Melbourne; London: Scribe.

Wilton, R. (1998). The constitution of difference: space and psyche in landscapes of exclusion. *Geoforum*, 29, 73–85.

Yack, B. (2012). *Nationalism and the moral psychology of community*. Chicago: University of Chicago Press.

Zimmerbauer, K. (2011). From image to identity: building regions by place promotion. *European Planning Studies*, 19, 243–260.

Zimmerbauer, K. & A. Paasi (2013). When old and new regionalism collide: deinstitutionalization of regions and resistance identity in municipality amalgamations. *Journal of Rural Studies*, 30, 31–40.

Zimmerbauer, K., T. Suutari & A. Saartenoja (2012). Resistance to the deinstitutionalization of a region: borders, identity and activism in a municipality merger. *Geoforum*, 43, 1065–1075.

3 Measuring local identity

The next chapters give a detailed analysis of the importance and political role of spatial identities for local communities in two Dutch municipalities. These two municipalities were selected since they were of roughly the same size, and had experienced an amalgamation in the last decade, but were not currently engaged in an amalgamation process. The differences between the more peripheral rural communities on Goeree-Overflakkee (see Figure 3.1) and the urbanised rural communities in Katwijk (see Figure 3.2), together with the diversity of the local communities within these amalgamated municipalities, were helpful in making useful comparisons of the different use of local identities during and after municipal amalgamations. These two municipalities were chosen in consultation with the Dutch Ministry of the Interior, which commissioned this explorative research, the results from which largely form the basis for the analysis of these case studies (Terlouw & Hogenstijn 2015).

Goeree-Overflakkee is a rural island with 14 villages some 30 kilometres to the southwest of Rotterdam. After several failed attempts for cooperation between municipalities on the island, the central government forced the municipalities on Goeree-Overflakkee to merge in 2013. This was strongly opposed by large sections of the population in the old municipality of Goedereede, who feared that their local identity would be threatened by an amalgamation.

Katwijk, the subject of the other case study, is much more urbanised. The municipality of Katwijk incorporates four villages located close to the medium-sized town of Leiden, some 30 kilometres to the southwest of Amsterdam. It used to be part of a very strong regional cooperation, which has weakened over the last decade. It was created when, in 2006, three municipalities voluntarily merged. The protection of the different local identities was an important topic in this amalgamation process.

Almost all other studies on the importance of local identities focus on the extent to which individuals identify themselves with the local level.

This is mostly done by comparing local identification to the identification with other spatial scales, such as the regional, the national and sometimes the European level (Kato 2011; Casey 2010; Jones & Desforges 2003; Antonsich 2007, 2010a, 2010b, 2011; Soguel & Siberstein 2015; Brown & Deem 2016; Moreno 2007). This conceptualisation of local identities as a single, numerical variable indicating the degree to which individuals identify locally is very different from how spatial identities were discussed in Chapter 2. By focussing on individual identity as a numerical variable, the characteristics of local identities and the reasons why they are important for local communities are neglected. This is the focus of this book, which therefore studies spatial identities in a different way.

While identities are not part and parcel of everyday life, they are often created during an interview. Answers are commonly determined more by the questions than by the identity of the interviewee (Fern 2001, 145). Local identities are, however, not constructed by isolated individuals, but are part of collective discourses. Local and other spatial identities are, as was discussed in Chapter 2, not spatial facts, but social constructs. They are created and reproduced through discourses by stakeholders. They materialise in, for example, planning documents, newspaper reports or websites (Paasi 1991, 2002, 2010, 2011, 2012).

> Collective identity is not "out there", waiting to be discovered. What is "out there" is identity discourse on the part of political leaders, intellectuals and countless others, who engage in the process of constructing, negotiating, manipulating or affirming a response to the demand (. . .) for a collective image.
>
> (McSweeny 1999, 77–78)

Local identities are not tangible and fixed spatial facts, but are socially constructed in identity discourses which are disputed, reinterpreted and thus transformed (Paasi 2012, 3).

Our research on the role of identities in local politics started by reconstructing the existing dominant local identity discourses present in policy documents, party manifestos, local media and books about local history and daily life. This was done at different scale levels, including from the villages, the old-premerger municipalities (which sometimes coincide), the amalgamated municipality and its regional environment. The different dominant spatial identity discourses which emerged from the analysis of these sources were tested and refined in open interviews with five local key actors in each municipality. These were local politicians and leaders of local organisations which we had identified in the study of the local sources on identity discourses. The next sections discuss first some general key characteristics of Katwijk and Goeree-Overflakkee and give subsequently an overview of the different dominant spatial identity discourses in these two municipalities based on studying these local sources.

3.1 Goeree-Overflakkee

Goeree-Overflakkee is located a few dozen kilometres to the southwest of Rotterdam, which has the main sea harbour of the Netherlands. In 1751, it became an island when the islands of Goeree and Overflakkee were connected by a dam built and financed by the province of Holland to improve flood protection and land reclamation. Goeree-Overflakkee was part of a flourishing commercial agricultural system based on grain production on Overflakkee and commercial shipping which was dominated by the harbour town of Goedereede. This regional economic system declined during the nineteenth century, which transformed the island into an isolated peripheral rural region. It suffered severely from the flood disaster which hit the southwestern part of the Netherlands in 1953. The flood killed 488 of its inhabitants, predominantly in the southeastern part of the island. Seven thousand houses were damaged or destroyed and flooding of most of the island forced the evacuation of 60 per cent of its 33,000 inhabitants.

Afterwards, the national government invested heavily in a special programme to protect the Dutch delta against further flooding. Many dams were built, which also transformed the isolated position of Goeree-

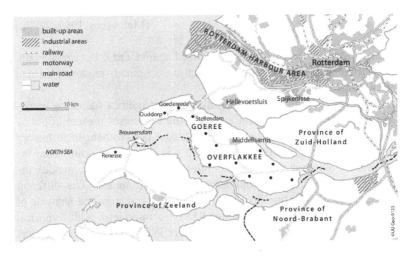

Figure 3.1 Goeree-Overflakkee: an island in the Dutch delta.

Overflakkee. In combination with national policies to stimulate economic development, this traditional, isolated agricultural region became influenced by the urban core area of the Netherlands. In the 1960s, it was opened up for tourism and a large part of its population started commuting to the then expanding Rotterdam harbour area. The subsequent decline and mechanisation of the harbour now affects employment opportunities of the islanders. Combined with the growing obsolescence of the tourist sector on the island, which is still largely based on campsites constructed in the 1960s, this resulted in a growing out-migration from the island. These regional economic problems are the focus of the first dominant identity discourse we identified by studying documents and interviewing key local actors (see Terlouw & Hogenstijn 2015 for a more detailed discussion of the documents analysed).

3.1.1 Dynamic island in the Delta

To reverse this decline, policy makers and entrepreneurs have produced numerous plans over the last decade to improve the prospects of the island. These have mostly been linked to the political debates which resulted in the municipal amalgamation in 2013. These stakeholders want to profit from its location in the Dutch delta, its proximity to large cities and its residential qualities based on its attractive rural landscape, its coastline and its recreational possibilities. Their rural island identity is used to promote it as an attractive residential area with an attractive island landscape. New inhabitants are necessary for the island so that it will have a large enough support base, especially for health services. The identity of the island is also linked to its potential for sustainable development. Innovative and locally rooted energy production is linked to the key qualities of this windswept island, surrounded by sea arms with a potential to generate tidal energy. These and other new types of economic activities will make the island a more dynamic and attractive part of the Dutch delta. This identity discourse identifies the island as a dynamic part the Dutch delta in which it has a development potential based on both its rural qualities and its innovative capacity.

3.1.2 Unspoiled tranquil holiday island

The second dominant identity discourse on Goeree-Overflakkee represents it as an unspoiled tranquil holiday island. It focusses on the position and attractiveness of the island for beach and water tourism, which have been an important but declining economic sector on this island since the 1960s. The expressed dynamic island identity of the first identity discourse, and especially the windmills used for sustainable energy production, conflicts with the quiet open coastal landscape and rural villages valued in this identity discourse. While the first identity discourse focusses on the links of the island with the urbanised north, this identity discourse focusses more on the similarities with other touristic islands to the south.

3.1.3 Traditional local communities

The third identity discourse focusses on the traditional character of the local communities on the island. The focus is on the historically grown social structures which value hard work, self-reliance and solidarity, partially based on a Christian heritage. It is linked to the historical isolation of the island and the harsh struggle for existence on this secluded island where people depended on each other. Although these elements are in decline, this discourse on traditional village identity is present to differing degrees in the local identity discourses of all the local communities on the island.

3.1.4 Unique Goeree

The fourth identity discourse focusses on the uniqueness of Goeree and its difference from the rest of the island in mentality and landscape. They claim that they have over time developed a more entrepreneurial mentality to cope with their frugal agricultural possibilities behind the dunes, compared to the wait-and-see mentality of the large farmers on the eastern part of the island, who used to profit from their fertile clay polders and their numerous cheap wage labourers. Their entrepreneurial mentality also made them successful in the tourist sector. Their virtuousness compared to the rest of the island is also reflected in their greater religious devotion and their financial prudence.

3.2 Katwijk

The municipality of Katwijk is located on the edge of the Dutch urbanised core area. Its coastal location gives it a somewhat marginal position against the large Dutch cities inland. Historically, its position was more important. Located at the mouth of the River Rhine, it had an important Roman presence. In the Middle Ages, it included the sacred place where the investiture of the counts of Holland took place. Later, especially during the seventeenth century, its position was overshadowed by the wealthy commercial cities in Holland, like Amsterdam, Leiden and Haarlem. Agriculture in the villages which now make up the municipality of Katwijk specialised in catering for specific urban demands. Katwijk aan Zee

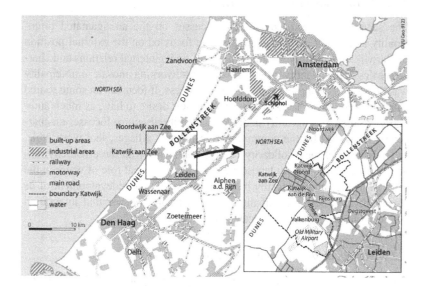

Figure 3.2 Katwijk: on the edge of the Dutch urban core.

specialised in fishing; the horticulturalist in Katwijk aan de Rijn concentrated on carrots and onions; those in Rijnsburg focussed on flower cultivation; and in Valkenburg they had an important horse market, which they boast is the oldest in the Netherlands. Since the 1960s, this relatively open rural area has been subjected to increasing urbanisation pressures from the Dutch urbanised core area. The present municipality of Katwijk was formed in 2006 through an amalgamation of the old municipality of Katwijk, which consisted of the two separate and distinct villages of Katwijk aan Zee and Katwijk aan de Rijn, the municipality of Rijnsburg and the municipality of Valkenburg. These four villages, in an amalgamated municipality surrounded by an urban context, are important elements in all local identity discourses.

3.2.1 Katwijk distinct from its region

The identity discourse which positions Katwijk in its regional context focusses on protecting its autonomy by selectively cooperating with many different groups of municipalities. Since amalgamation, Katwijk has focussed on protecting its independence from its neighbouring rural and urban municipalities. They pragmatically cooperate out of necessity, but regard too close regional cooperation as a threat to their autonomy.

3.2.2 Hardworking mosaic municipality

The second identity discourse also focusses on the amalgamated muni-cipality as a whole. Whereas the first one focussed on the external position of Katwijk, this identity discourse focusses on its internal relations and characteristics. This identity discourse of a hardworking mosaic municipality was developed during the amalgamation process. It focusses on some shared characteristics of the four villages and on their desire to have as much autonomy as possible. This unity in diversity identity discourse focusses on characteristics shared by the communities in all four villages, like work ethic, self-reliance and local solidarity. This is contrasted with the urban environment and linked to a rural landscape, with the important role of water in the form of the sea and especially the Rhine. Through the redevelopment of its riverbanks, the new municipality hopes to create some spatial integration between the four villages all located along this river.

3.2.3 Strong, distinct local communities

The four strong, cohesive and distinct local communities also generate identity discourses which focus on their differences instead of their similarities. These differences have been institutionalised through the establishment of neighbourhood councils during the amalgamation process, which were established with the explicit goal of protecting the different local identities of the communities.

3.2.3.1 Katwijk aan Zee: traditional Christian fishing village

One identity discourse focusses on Katwijk aan Zee as a Christian fishing village. This identity discourse is very traditional and is focussed on its glorious days as a poor but noble fishing village. Many native villagers are still proud of their own local dialect and many of them still have a traditional nickname. The old white church on the boulevard is a strong icon of this village. For a long time, it was a beacon for their fishermen and also signifies the dominance of traditional Christians in their local community.

3.2.3.2 Katwijk aan Zee: seaside family resort

Another identity discourse portrays Katwijk aan Zee as a seaside family resort. Its more tranquil character is contrasted with the mass tourism in neighbouring seaside resorts like Noordwijk and Zandvoort. This identity discourse focussing on a traditional family type of tourism tries to align itself as much as possible with the dominant identity discourse of Katwijk aan Zee as a Christian fishing village.

3.2.3.3 Katwijk Noord: urban tower blocks

These traditional local identity discourses on Katwijk aan Zee are in sharp contrast with the identity discourse on Katwijk Noord as an out-of-place urban neighbourhood. This identity discourses highlights the large, imposing, anonymous tower blocks which are connected with many urban social problems like drugs abuse and migration. It is represented as an urban neighbourhood which does not fit in with the rest of the municipality and with problems that should be addressed.

3.2.3.4 Katwijk aan de Rijn: rural village

The identity discourse on Katwijk aan de Rijn presents it as distinct from the larger fishing village of Katwijk aan de Zee. This is linked to their different horticulturalist past, but much attention is given to their strong local associations and their well-attended local festivals. They are considered to be easy-going folks who can easily associate with others.

3.2.3.5 Rijnsburg: flower traders

The flower cultivation and trade dominates the identity discourse of Rijnsburg. Their historically developed commercial attitude is linked to their economic successes and their tendency to show off their wealth. Their shared focus on the flower trade creates strong ties in this village. They are proud of the redevelopment of their old town hall into a much-used and valued community centre.

3.2.3.6 Valkenburg: village between cities

The smallest village in the amalgamated municipality of Katwijk is Valkenburg. Its dominant identity discourse stresses its traditional, close community life, which is very distinct from the neighbouring city of Leiden. This is exemplified by their festivities around their traditional horse fair, which they boast as being the oldest one in the Netherlands. This tightly knit village community is contrasted with a recently built nearby neighbourhood and an even bigger planned new neighbourhood.

3.3 Interviews

All these different local identity discourses, summarised in Table 3.1, were first validated and refined in explorative interviews with key local actors. The different local identity discourses were then used as the basis of a

Table 3.1 The dominant spatial identity discourses on Goeree-Overflakkee and in Katwijk

Section		Scale	Main promoters	Pictures
§3.1	Goeree-Overflakkee			
§3.1.1	Dynamic island in the Dutch delta	Region	Policy makers Entrepreneurs	Island in the Dutch delta Windmills along dike
§3.1.2	Unspoiled tranquil holiday island	Municipality	Municipality Tourist industry	Marina Beach
§3.1.3	Traditional local communities	Villages	Local community	Old church New church
§3.1.4	Unique Goeree	Village	Inhabitants of Goedereede and Ouddorp	Historical centre of Goedereede Placard against amalgamation
§3.2	Katwijk			
§3.2.1	Katwijk distinct from its region	Region	Policy makers	Logo Greenport Map with HollandRijnland
§3.2.2	Hardworking mosaic municipality	Municipality	Municipality	Mosaic Hardworking footballer
§3.2.3	Strong, distinct local communities	Villages	Local community	
§3.2.3.1	Katwijk aan Zee: traditional Christian fishing village	Village	Local community	Old white beach church Street art
§3.2.3.2	Katwijk aan Zee: seaside family resort	Village	Municipality Tourist industry	Beach cabins Fishing days
§3.2.3.3	Katwijk Noord: urban tower blocks	Village	Other inhabitants of Katwijk	Tower blocks
§3.2.3.4	Katwijk aan de Rijn: rural village	Village	Local community	Old windmill
§3.2.3.5	Rijnsburg: flower traders	Village	Local community	Flower parade Community centre
§3.2.3.6	Valkenburg: village between cities	Village	Local community	Centre of old village New housing estate

Source: author's own data.

topic list we used in the semi-structured interviews. Each identity discourse was illustrated by two pictures, many of which we came across in the analysis of the documents related to local identities. We decided to use pictures in the interviews since pictures are suitable tools for discussing abstract concepts like identity in an open manner (Croes *et al.* 2013; Holgate *et al.* 2012; Müller 2011). The interviews always started with an open question on what characterises their local identity. Then the different identity discourses were discussed by showing the pictures, which sometimes needed further explanation. Later on, the changes in the character and use of local identity, its relation with regional identity, conflicts related to identity and their views on expected future changes were discussed (for the complete list, see Terlouw & Hogenstijn 2015, 165–168). Towards the end of the interview, the interviewees were asked if they missed some aspects of local identity. Most interviewees felt that we covered all relevant topics related to local identity. Some mentioned other aspects. These were generally mentioned once. Only the influence of agriculture and its diversity was mentioned by several interviewees on Goeree-Overflakkee. All interviewees were given the opportunity to discuss these aspects, which were included in the analyses of the texts of the interviews.

Based on the analysis of local sources and the interviews with local key actors, we identified and approached a wide range of active members within the local communities. We interviewed 58 people in the final months of 2014. During that time, the interviewer (Maarten Hogenstijn) stayed for one week in both municipalities and conducted an additional 22 street interviews. We stopped contacting new informants once the information generated from interviews converged and no new themes or opinions were raised. The principal researcher was present during a quarter of the interviews and listened to all recorded interviews. These were transcribed in 587 pages. In a first analysis of the transcripts, we identified 189 relevant topics. These were further systemised and condensed into 29 topics, which form the basis of the following chapters (Terlouw & Hogenstijn 2015).

References

Antonsich, M. (2007). *Territory and identity in the age of globalization: the case of Western Europe* (PhD thesis, University of Colorado).

Antonsich, M. (2010a). Exploring the correspondence between regional forms of governance and regional identity: the case of Western Europe. *European Urban and Regional Studies*, 17, 261–271.

Antonsich, M. (2010b). Meanings of place and aspects of the self: an interdisciplinary and empirical account. *GeoJournal*, 75, 199–132.

Antonsich, M. (2011). Grounding theories of place and globalisation. *TESG*, 102, 331–345.

Brown, A.J. & J. Deem (2016). A tale of two regionalisms: improving the measurement of regionalism in Australia and beyond. *Regional Studies*, 50, 1154–1169.

Casey, R. (2010). Community, difference and identity: the case of the Irish in Sheffield. *Irish Geography*, 43, 211–232.

Croes, R., S.H. Lee & E.D. Olson (2013). Authenticity in tourism in small island destinations: a local perspective. *Journal of Tourism and Cultural Change*, 11, 1–20.

Fern, E.F. (2001). *Advanced focus group research*. London: Sage.

Holgate, J., J. Keles & L. Kumarappan (2012). Visualizing "community": an experiment in participatory photography among Kurdish diasporic workers in London. *The Sociological Review*, 60, 312–332.

Jones, R. & L. Desforges (2003). Localities and the reproduction of Welsh nationalism. *Political Geography*, 22, 271–293.

Kato, Y. (2011). Coming of age in the bubble: suburban adolescents' use of a spatial metaphor as a symbolic boundary. *Symbolic Interaction*, 34, 244–264.

McSweeny, B. (1999). *Security, identity and interests*. Cambridge: Cambridge University Press.

Moreno, L. (2007). Identités duales et nations sans état (la question Moreno). *Revue Internationale de Politique Comparee*, 14, 497–513.

Müller, F. (2011). Urban alchemy: performing urban cosmopolitanism in London and Amsterdam. *Urban Studies*, 48, 3415–3431.

Paasi, A. (1991). Deconstructing regions: notes on the scales of spatial life. *Environment and Planning A*, 23, 239–254.

Paasi, A. (2002). Bounded spaces in the mobile world: deconstructing "regional identity". *TESG*, 93, 137–148.

Paasi, A. (2010). Regions are social constructs, but who or what "constructs" them? Agency in question. *Environment and Planning A*, 42, 2296–2301.

Paasi, A. (2011). The region, identity, and power. *Procedia Social and Behavioral Sciences*, 14, 9–16.

Paasi, A. (2012). Regional planning and the mobilization of "regional identity": from bounded spaces to relational complexity. *Regional Studies*, 47, 1206–1219.

Soguel, N. & J. Silberstein (2015). Welfare loss with municipal amalgamations and the willingness-to-pay for the municipality name. *Local Government Studies*, 41, 977–996.

Terlouw, K. & M. Hogenstijn (2015). *"Eerst waren we gewoon wij en nu is het wij en zij": gebruik slijtage en vernieuwing van regionale identiteiten*. Den Haag: Ministerie van Binnenlandse Zaken en Koninkrijksrelaties. www.rijksoverheid.nl/bestanden/documenten-en-publicaties/rapporten/2015/05/01/onderzoeksrapport-over-lokale-en-regionale-identiteiten/eerst-waren-we-gewoon-wij-terlouw-hogenstijn-2015.pdf.

4 Local identities analysed
Change for the better or the worse

This chapter discusses some general characteristics of local identities as they are perceived by our interviewees in both municipalities. The subsequent chapters then analyse how local identities are used in the local politics of Goeree-Overflakkee (Chapter 5) and in Katwijk (Chapter 6). This chapter shows that in the local population there is a widespread agreement on what characterises their local identity, but that they differ widely in their valuation of this identity. This positive or negative opinion influences their behaviour towards their local identity; many want to protect it, while others want to change it. This is an important driving force in local politics, which becomes even more important, as we will see in Chapters 5 and 6, when the different ways in which local identities are used during and after municipal amalgamations are analysed.

4.1 Local identity and community values

Traditional qualities like hard work, modesty, solidarity, self-reliance, adaptability and resilience dominate the answers our interviewees gave to the opening question on the general characteristics of their local identity. The index to this book lists a dozen of these different but related community values. On this general level, the local identities of all the different local communities are based on quite similar social characteristics. Our interviewees base these answers largely on their daily life in the local community. These social characteristics of their local identity are frequently linked to a traditional way of life, customs, local events and the role and character of voluntary associations. Although very similar traditional social characteristics were used to characterise all local identities, they were mostly linked to their own distinct local community. This was especially the case when we confronted our interviewees with pictures representing the different local identity discourses.

Figure 4.1 Street painting at the seaside in Katwijk.

Source: photo taken by Maarten Hogenstijn.

In all documents we studied, the identity of Katwijk aan Zee is linked to its origin as a fishing village. One of the pictures we used in the interviews shows a street painting on the seaside boulevard of Katwijk aan Zee depicting traditional fishing, the old church and seaside tourism (see Figure 4.1). Most of our interviewees reacted to this picture by commenting on the role of fishing for the local identity of Katwijk aan Zee. Fishing and the traditional harsh way of life are often mentioned as the origin of the distinct local identity in Katwijk aan Zee. This is linked to the characteristics of a tightly knit, traditional local community, like solidarity, self-reliance, resilience and social control. It is also frequently the starting point of making comparisons across time and space. Most interviewees discuss the decline of the tightly knit fishing community and the shrinking differences between Katwijk aan Zee and the other villages in the municipality of Katwijk.

This loss of difference is a recurrent theme when people reflect on the local identities in their municipality. The loss of local identity is an important theme in most of our interviews. The general theme of loss of local identity is the backdrop of many of the topics discussed in this chapter. In §4.9 and §4.10, its general importance in spatial identity discourses will be discussed more systematically.

The diminishing role of fishing for the livelihood of the local community in Katwijk aan Zee is frequently mentioned, but the continued interest and identification with the fishing tradition is also a much used narrative. For instance, an exposition in the local museum on fishing and the daily life of the fishers before 1960 attracted many visitors. However, the follow-up exposition on the techniques of modern fishing attracted far fewer people. Some regard fishing as a relic of the past which clouds the development of a vision on the future and hinders the economic policies of the municipality. A local politician commented on the picture in Figure 4.1:

It is deliberately conserved during festivals when those old ladies put on their folklore costumes and repair fishing nets all day long. Though I think this is no longer how it is, this no longer characterises Katwijk. (…) You have also to look to the future, there is more than this traditional image.

The fishing tradition is linked to the continued importance of values related to self-reliance in the provision of livelihood and based on a shared mentality of hard work within the local community. But these characteristics are commonly seen, even by inhabitants of Katwijk aan Zee, as a general characteristic of all the local communities in the municipality of Katwijk. These general traits are linked to the different forms of livelihoods in the different villages. For Rijnsburg, hard work is linked to their strong entrepreneurial spirit. The different ways in which these similar qualities are used are also linked to differences in the local identities of the villages. For instance, some businessmen operating in the municipality of Katwijk contrast the characteristics of the cautious, inwardly oriented inhabitants of the old fishing village Katwijk aan Zee with the people living in the inland village of Rijnsburg, who work hard selling their flowers:

That is because they trade originally with Germany, but now of course with the entire world. That is why they are very outwardly oriented. And impertinent. Very direct. It is not like: "Sir, could you arrange this or that?", but: "Darn it! What the heck is going on!" That's how these folks talk!

These traditional roots of local identities are used not only to differentiate in space between the different local communities, but also to comment on the development over time. "It is a concern for the future that the traditional fishing in Katwijk and flower trade in Rijnsburg are threatened" (local community worker). The reliance on locally rooted traditional economic activities is seen by many as a weakness now the transformation of the economy is gathering pace.

4.2 Heritage and the different views on the future

The history of this church (Figure 4.2) reflects the history of Katwijk aan Zee. This became apparent in our study of the identity-related documents. The role of this church in local history was, however, hardly mentioned by our interviewees, who focus, as we will see later, on its current role in the local community and identity discourses.

The first church built on this location was destroyed in 1572 by the Spaniards during the Dutch Revolt. The present church was built in 1640. Through the gradual process of coastal erosion, it now stands at the seafront. Its tower was an important beacon for the fishermen of Katwijk. This white church is also depicted in many romantic paintings of the late nineteenth century as the background of the fishing boats on the beach from which fishermen and women bring fish ashore. In 1885, the parish built a newer and larger church and sold the old church to a ship owner who used it for storage. In 1924, it was again transformed into a church for a newly formed, more orthodox-protestant parish. During the Second World War, its tower was partly destroyed by the

Figure 4.2 The old church in Katwijk.
Source: photo taken by Maarten Hogenstijn.

German army. The church is frequently mentioned and depicted in the documents we studied on the local identity of Katwijk. The picture in Figure 4.2 is linked by our interviewees with the identity of Katwijk. "The white church and fishing is the basis of what Katwijk is" (local politician). It is sometimes mentioned in our interviews that, when people from Katwijk aan Zee come back from holidays, many of them first drive along the boulevard and pass the church in order to feel at home again. It is also widely used in folders to attract tourists. Many find that it communicates a stereotypical image of Katwijk. They sometimes comment on its use by the Dutch media as the background in TV reports on incidents like the high levels of drug abuse among the youth in Katwijk. This church is then used in reports which portray Katwijk as a whole as a backward, traditional Christian municipality.

Our interviewees mostly focus their discussion on what is allowed and what should be allowed in the church square. Orthodox Christians want to keep it tranquil and empty of new elements, and perceive that it is threatened by disrespectful tourists, especially on Sundays. One elderly local administrator, born in Katwijk aan Zee and with a Christian background, commented on the picture of the church with the new square with bronze fish sculptures: "It doesn't have to be like this! No, this is of course too much embellishment for such a respectable and frugal village. For me this is very awful ...". Others agree with him on the dominant orthodox Christian character of their local community, but they want to change it. In their opinion, the image of the church is overused in the external communications of the municipality and entrepreneurs, which in their view unfortunately strengthens the stereotypical image of Katwijk aan Zee in the outside world. "Ah, the white church. That is very stereotypical Katwijk. This is really how Katwijk is promoted, with all other things, the beach, the fishing. I doubt that is accurate" (inhabitant who works outside the village). Some want this coastal church square to be developed with sidewalk cafés and bars in order to attract more tourists. One local politician born in Katwijk aan Zee, but who has lived many years in other places in the Netherlands, comments:

> This is magnificent. (...) Also the fountain with that wave. It is similar to other places, but then you would also have pubs and terraces. In Brabant you would go from the church into the pub. That is impossible here. And then I think we waste opportunities. When it's hot, it would be so nice to sit here. (...) We really let opportunities pass. Katwijk is so much more than going to church. Count them all together and they still are not the majority. That is an inconvenient truth, I respect them, but ... they should give others also leeway.

This case of the old church in Katwijk aan Zee illustrates that in general our interviewees agree to a remarkable degree on the main characteristics of the local identity, but differ considerably in how they value it and what actions should be undertaken to protect or change it. Different groups within a local community mostly agree on what characterises the identity of the local community. This cognitive consensus frequently coincides with affective disagreements on how to value this local identity, and political disputes on what action (conation) is necessary to protect or change it.

In Katwijk, dealing with different valuations of the dominant local identity has become part of a shared secondary identity within the municipality (see also §2.9 on the relation between primary and secondary identities). For instance, the general characterisation of the local identities in Katwijk hardly differed between elderly orthodox-protestant members of the church council and local youths of the Gothic music scene. These youths we interviewed were perhaps even more aware of the local differences, while they knew where in the municipality their appearance, with white crosses on their black clothes, elicited the most resistance. The gothic youths in Katwijk do not like the sabbatical rest on Sundays, but have learned to deal with it by cycling to the pubs in nearby towns.

> Why would you change it, that would result in even more trouble with that group who dislikes it. That is in my opinion a bit useless for this. If you want to do something on a Sunday you cycle to Leiden or Noordwijk.

Many of these youths also consider moving towards the nearby cities of Leiden or Amsterdam.

The different local groups are very aware of what local political issues are very sensitive for the other groups. They mostly avoid confrontations on the most sensitive issues. This is a clear example of how differences about primary identities are accommodated by a secondary identity, as was discussed in §2.9. The role of secondary identities are also discussed in the next chapter, in §5.5, when the political confrontations about Sunday rest in the recently amalgamated municipality Goeree-Overflakkee are analysed.

4.3 Types of identification

Different kinds of people value the same local identity in different ways. The quotes discussed above suggest that this differs between the elderly who have lived all their life in a community and the more mobile and especially younger inhabitants. We found similar differences between the local identification of different kinds of people in many of our interviews in both Katwijk and Goeree-Overflakkee. Three main groups stand out: the natives, the migrants and the young. These differences are very similar to the different types of place attachments identified by David Hummon (1992) and further elaborated by Maria Lewicka (2011a; 2011b). These environmental psychologists study the different ways in which inhabitants attach or detach themselves to the local community in the place where they live. Contrary to many others, they do not measure simply the degree to which individuals identify with their local community. Identity is a complex phenomenon which cannot be captured by a number. Instead they focus on the different ways in which people position themselves towards their local community. Based on their work and the results of the interviews, it is useful to make a distinction between the traditionally rooted, the actively attached and the locally indifferent.

4.3.1 Traditionally rooted

The traditionally rooted identify locally based on their long residence. They have strong social relations within the local community based on their family ties and long-term friendships. They are part of a strong social network within the local community, but have few meaningful social relations with people elsewhere. Their relations focus on informal local social networks, but they participate much less in more formal political organisations. Their relations bond more locally rather than bridging to the outside world. They are typically, but not exclusively, born locally and have always lived there. Compared to the others groups, they are less educated, older and more conservative (Lewicka 2011a, 2011b).

Their attitude towards local identity can be illustrated by quotes from our interviews. "I was indeed born and bred in Katwijk. I was born just behind the boulevard. My parents had a grocery so I know everybody" (middle-aged woman who lived her whole life in Katwijk). A very active local politician who migrated to Goeree-Overflakkee 25 years ago and who is considering moving back after his retirement comments: "Those who are born and bred here have an inward orientation. It is a bit, like knows like."

4.3.2 Actively attached

The actively attached also identify locally, but this is not built on traditional roots but on choice. Their local identity is not instinctive and taken for granted, but based on conscious decisions. They have deliberately chosen their place of residence. Most were born elsewhere, but some were born locally, moved away and made a conscious decision in later life to move back. Their choice of where to live was mostly based on a comparison with other possible places. Moving here is based on their positive valuation of the local community and its identity.

The actively attached develop many local contacts, but retain many of their contacts with people and places elsewhere. Whereas the traditionally rooted bond together more, the actively attached make more bridges to the outside world. They identify not only locally, but also sometimes with other places and especially with the wider region. While the traditionally rooted identify locally primarily based on social relations, the actively attached value more spatial elements like landscape and heritage sites. These more regional characteristics are important elements for the quality of life in their chosen place of residence. Their local identity is less inward orientated, and is more layered and linked to other spaces. The actively attached are mostly middle-aged, more educated, very interested in local and regional history, value the conservation of local and regional heritage and landscape, and are active in more formal associations and local political organisations (Lewicka 2011a, 2011b).

Their attachment to their chosen home is frequently mentioned by the immigrants we interviewed. A middle-aged migrant explains why he moved to Goeree-Overflakkee.

I come from across the water and have chosen consciously to live here. I lived in Spijkenisse and worked in the Rotterdam harbour, but disliked it. In Middelharnis we had everything close at hand for our children, a swimming pool and secondary schools. What is especially attractive is that everybody feels responsible and bonded. It is really an island, the sense of togetherness on one island.

An active migrant comments: "People who came here from across the water are more actively propagating the island than the islanders themselves, they take everything for granted." "Members of village councils are often not the people who were born and bred here."

4.3.3 Locally indifferent

Not everybody feels attached to their local community. Some are so alienated that they have a negative attitude towards their place of residence. They are mostly dissatisfied with society in general and have few social contacts within or outside the local community. Others, the locally indifferent, are not so much alienated, but are more uninterested in their local community. They identify less with spatial communities and more with other types of communities. The young especially develop more social contacts outside their local community. The locally indifferent focus their identification outwardly. They develop social relations with other individuals with whom they share a similar lifestyle. This kind of local indifference was expressed especially by the youths we interviewed. "Goeree-Overflakkee is dull, nothing much ever happens here, and everybody watches you! I really don't want to stay here. I want to live in Rotterdam. You cannot go out here, nothing. Goeree-Overflakkee is an island with nagging old people."

Different groups of people thus link their individual identity in very different ways with their local community. This partly explains why, though they cognitively largely share the dominant local identity discourses, they value these very differently and react to them differently in their behaviour. People want to either preserve, change or avoid the elements of the dominant local identity discourse. This resembles the classical distinction made by Albert Hirschman (1970) between the different ways in which people react to a situation they dislike through voice, exit or loyalty. They can voice their opposition by complaining and trying to change the situation. People can exit from the situation they dislike through, for instance, migration or by disengaging oneself from the local community. Or one can be loyal and go along with the situation while fearing the social or economic costs of either voice or exit.

4.4 Established local communities and migration

While natives tend to be traditionally rooted, migrants are mostly actively attached or locally indifferent. These differences in how people identify locally are also linked to the relations between natives and migrants. Migration is an important topic in our interviews on the role of local identities for local communities.

Moving between local communities makes people aware of local differences and the social consequences of migration. Somebody who moved between neighbouring local communities comments on the naturalness of the differences between local identities: "Look, it has always been so, I come from the other part of the island and then they say: 'Oh yes, those Ouddorpers' and then they say here: 'Oh yes, those Flakkeeënaars'." A member of a village council on Goeree-Overflakkee commented on the established attitude of islanders towards people from the outside: "Everything that came from outside was bad. Migrants are still branded as people coming from the other side of the water. If you are not born and bred here, you are not really included in the community."

Migration is an important element in the construction and use of local identities. The experiences with migration influence how different people identify locally. The traditionally rooted, who are mostly born and bred locally and have stayed in their local community, are confronted with migrants moving in and out of their community. Those moving out are more locally indifferent or detached youths, who look for another, predominantly more urban, residence which better fits their chosen lifestyle. Those moving in do this mostly based on a deliberated decision. These predominantly middle-aged migrants are attracted to living in this place. The physical characteristics of a place are especially important for them. The natives in the local community are more indifferent to this. The local identity is also, for many, an important reason to migrate from or to a specific place. Others who were initially more locally indifferent become, over time, more actively attached to their local community.

The relation between migrants and natives is frequently mentioned by our interviewees. For instance, a member of a village council on Goeree-Overflakkee who moved to the village almost 40 years ago because of professional opportunities reacted to the opening question on the general characteristics of the local identity as follows:

> There moved in a lot of new people. Those are called here "people from the other side". Those are actually for the natives a kind of second-class citizen. The island culture was very strong when we came here in 1977. It was like you may take part while they needed us, but don't think you can make decisions. Then it was stop, you don't come from here and don't think you can change us. That has diminished slowly but surely, while ever more people moved in from the other side of the water. What is accomplished in this village is to their irritation mostly done by those "people from the other side". The natives are critical, they are a bit afraid of those strange people from the other side of the water. They don't belong here with us. That is still important. If they just a little bit think you want to control things, you get a lot of criticism. (...) But it is very difficult to get people to participate. There are a few, but mostly incomers, who say yes we have to do it together. But there are three churches with many members, but a community spirit of resolving things together, that is still problematic. But those are the downsides, but there are of course a lot of upsides to this island. It is naturally such a beautiful island. There is of course a split between both parts, and the western part distrust this part, while this part less so. But you live all on the island Goeree-Overflakkee. If we travel back from Rotterdam we say: "We are back on the island." That grows over time. We are very happy here, we don't want to leave. There are more people who came here for work and originally wanted to leave after retirement. But I have gradually grown so attached to the island.

This quote also illustrates that newcomers who become actively attached to their place of residence tend to become more attached to its physical aspects, like the landscape in the region, than to the specific local community where they live. The actively attached identify in general more regionally, while the traditionally attached identify more locally.

The difficult relation between migrants and natives is also discussed in our interviews in Katwijk. In all four local communities in the amalgamated municipality of Katwijk, the difficulties of newcomers to be accepted as fully fledged members of the local community are mentioned. Newcomers are largely excluded from the solidarity of the local communities. They are considered as a danger to the moral order and it is feared that newcomers will undermine the power base of the established. They are branded as disrespectful to local customs and are feared to illegitimately take over power. This is neither a new nor a specifically rural phenomenon. In 1949, researchers investigating the changes in rural life in the peripheral eastern part of the Netherlands made very similar observations:

> The closed character of the hamlets is especially evident in their attitudes towards strangers. Before a new peasant can count himself as a group member, he has to comply with certain rules. (...) Has the stranger a lasting occupation in the hamlet, then it is up to him if he can be tolerated. Many have understood this and approached the peasant with respect. Then he can count on support from the community.
>
> (Groenman & Schreuder 1949, 87–88)

These tensions between the established and newcomers were also studied by Norbert Elias and John Scotson (1965). They analysed the relation between the inhabitants of older and newer working-class neighbourhoods in the industrial city of Leicester. The inhabitants of the older neighbourhood felt threatened by migrants living in a newly built housing estate. Despite having the same social background, these migrants were considered as morally inferior and transgressing the established customs of the local community. They were seen as a threat to the established local power structure dominated by the established. This perceived threat to their way of life strengthened the identity of the established community. Their moral superiority was contrasted to the bad characteristics and behaviour of the newcomers. The inferiority of the newcomers thus became an important element in the local identity discourse of the established. Their longer residential history enabled the established to develop a tight social network with distinct but informal standards of behaviour for dealing with each other and taking care of each other's interests and sensibilities. The newcomers were unaware of these informal rules and thus threatened the established community and its power structure. It was also difficult for the newcomers to develop their own social network while the established dominated the local community (Elias & Scotson 1965).

4.5 Scale strategies in local politics

These relations between established and outsiders frequently dominate local politics, but can change over time. For instance, the rivalry between established and outsiders in the neighbourhoods of Leicester disappeared a long time ago. The distinction between these neighbourhoods is no longer relevant for the local community. The memory of this rivalry is fading even for older residents (Hogenstijn & van Middelkoop 2008). The power balance between established and outsiders is not fixed, but can be changed, especially while the relations between groups within local communities are part of broader power structures in the wider world. Four different scale strategies can be discerned: excluding within the local scale (§4.5.1), mobilisation within the local scale (§4.5.2), upscaling the conflict (§4.5.3) and downscaling the conflict (§4.5.4).

4.5.1 Excluding within the local scale

Within a local community, a common strategy is to exclude others from influential positions. Newcomers are thus, as discussed above, commonly excluded socially and politically by the established (Hogenstijn *et al.* 2008).

4.5.2 Mobilisation within the local scale

Another strategy is to mobilise the members of a conflict group or expand the conflict group to include other groups. This strategy mostly involves the mobilisation of the locally indifferent (§4.3.3). People can lose their indifference especially when conflicts threaten individual interests like property or personal well-being. By claiming to defend their interests, the opposite parties in the local conflict can try to mobilise the locally indifferent. For instance, as we will analyse in more detail in the next chapter (§5.1), those opposing municipal amalgamation in Goedereede not only focussed on the threat to local identity in their communication with the traditionally rooted and the actively attached. They also tried to mobilise the locally indifferent by circulating frightening stories on higher local taxes and diminishing services in the amalgamated municipality.

Besides these two strategies implemented at the scale of the local community, there are also two strategies which can be used that cross this local scale. Groups can try to improve their position by changing the arena of the conflict. They can jump scale by ether upscaling (§4.5.3) or downscaling (§4.5.4) (Hogenstijn *et al.* 2008).

4.5.3 Upscaling the conflict

The upscaling of the conflict arena enables some groups to use their links with powerful groups or individuals at higher spatial scales to improve their position locally. In many cases, the conflict between different groups within a local community is quite similar to conflicts elsewhere, also involving actors operating at a higher spatial scale. If a group has a stronger position at higher spatial scales, it can try to use this position to augment their local power, and thus win the conflict locally. This upscaling is commonly used by a locally weak group. They try to shift the conflict to a higher spatial scale, where they have a stronger position on the power balance. In this way, the relevance of the local power sources of the other group diminishes. In particular, newcomers who are excluded from the local power structures by the established within the local community frequently use outside allies to improve their position. For instance, in discussions on nature and landscape preservation, local natives frequently lose out to migrants with better connections outside the municipality. One prominent community leader in Katwijk complains about how this blocks the improvement of a local road:

> Then they make a big fuss about a few plants, but the people who live there, this is my profound belief, live in a polluted environment, while from 15:30 till 19:00 there is one big traffic jam with running motors. That can't be healthy. But that guy from the environmental association continues to bugger us, and with his Natura2000 in his hand he wins in court, in some way or other.

We will also discuss in Chapter 5 that proponents and opponents of municipal amalgamation frequently use this strategy to find allies outside the local community.

4.5.4 Downscaling the conflict

Downscaling the conflict weakens the other group's position by restricting the conflict to a smaller area. A group might try to reframe the arena of the conflict by focussing on a small part of the conflict. They then can blame the conflict on nagging individuals who only complain because of their special interests. Labelling opponents as motivated just by a "Not In My Back Yard" (NIMBY) mentality is a very effective downscaling strategy. Also our interviewees frequently claim that opponents only act out of individual interest and neglect the collective interest of the community they claim to represent. A migrant to Goeree-Overflakkee who is actively attached to the island and is involved in projects to stimulate regional development brands his local political opponents in this manner: "NIMBY feeling is, as you know, always there. In part rooted in conservatism, in part lack of knowledge. They believe that if it worked out alright yesterday, it will work out tomorrow."

4.6 New migrants in old villages

Our interviewees agree that, although it was very difficult for migrants to integrate in the local community, this has become easier in recent years, mostly because of the continued inflow of migrants. The established have lost some power to the newcomers in the local community. The established especially lost influence in the larger amalgamated municipalities. The influence of the actively attached migrants has increased over time. They were attracted to specific aspects of the local identity in their place of choice. They have become actively attached to the regional landscape, the amenities and liveability of the local neighbourhood of their new home. The growing identification of migrants with especially "authentic" localities is frequently mirrored by a perceived decline of the traditional local identity by those born and bred there.

Figure 4.3 shows the old harbour in Goedereede. The picturesque old harbour is frequently mentioned in documents as very characteristic of the strong local identity of the old town. Our interviewees recognise it as typical for the traditional town. They focus, however, not on its past as an important old seaport with rich merchants dating back to the Dutch Golden Age. Our interviewees focus on its current inhabitants who migrated from the city and whose presence is linked to the demise of the traditional, tightly knit local community.

> The real Goereeër does not live here anymore. Especially around this harbour in the centre of the city of Goedereede live many who have migrated from the outside to the island. Thus the native Goederedenaar no longer lives there. But it used to be a true community in the past. I won't call them closed off, but a very close and caring community. Everybody visited everybody else during the day and drank coffee together, regardless if it was morning, afternoon or evening. It

Figure 4.3 The old harbour in Goedereede.
Source: photo taken by Maarten Hogenstijn.

was always cosy too. Everybody was ready to support if something
came up. That is diminishing, I have to say.

(Leading opponent of municipal amalgamation)

Migrants moving into old houses which are very characteristic of a place
is a very common phenomenon (Hogenstijn & van Middelkoop 2008).
Some old working-class areas in Katwijk are, however, still dominated by
those born and bred there. The fishing village of Katwijk aan Zee has
many neighbourhoods with rental houses built for fishermen families. A
local social worker comments:

> But if you are import, then it is not easy to become part of the com-
> munity. They have their own family, their own clique, and then a
> stranger arrives ... rather not, preferably someone from their own
> family, church or circle. Thus a stranger will much more feel at home
> and be much easier accepted in Katwijk Noord than in the old centre
> of Katwijk aan Zee.

4.7 Urban migrants in Katwijk Noord

The neighbourhood of Katwijk Noord was built in the late 1960s to alleviate the housing shortages in Katwijk aan Zee. It is located opposite to Katwijk aan Zee on the other side of the River Rhine. This river is traditionally nicknamed by many as the River Jordan, which symbolises the Christian character of these villages. This is in contrast with the nearby Dutch cities. Katwijk Noord was traditionally stigmatised as on the other side of the Jordan, as outside the promised land of Israel and inhabited by pagans, or in the words of a local resident, "foreigners and anti-social families".

> This is the other Katwijk. I deliberately call it the other Katwijk, while you find all kinds of trash, shit and squalor in these three tower blocks. You don't want to be found dead there. It is a kind of refuge for people who cannot find a place to live in other municipalities. Initially, it was to be a first place of residence, from which people could then move to a better place. Those who live there experience it differently, but in the local representation it is simply a place you don't want to be. It is somewhat the dregs of society. I frequently talk with residents, there is a lot of shit, but that has to do with migrants from Leiden with dubious backgrounds. You come across them everywhere, but they mass there, while there are openings. It is a part of Katwijk, as somebody told me, about which people living on the other side of the Jordan think they know all about. Across the water ... it is

Figure 4.4 Tower blocks in the neighbourhood of Katwijk Noord.
Source: photo taken by Maarten Hogenstijn.

now called Katwijk Noord, illustratively it used to be called Hoornes-Rijnsover, but that had a bad image. People labelled it as the neighbourhood with foreigners, the neighbourhood with problems. Two years ago they renamed it Katwijk Noord, with new signs to get rid of Hoornes-Rijnsover, even though that was well known.

This local politician continues in the interview to stress, like many other interviewees, that the inhabitants object to being so negatively labelled and that pictures of these tower blocks are always used in media reports on social unrest, housing problems or other negative developments. He ends his comments on this picture quite ominously. He recounts a recent conversation he had with an inhabitant of this neighbourhood.

He quite consciously talked about the other side of the Jordan, as a Gaza strip. I said, but then the whole lot will be blown up, including the mayor and all the municipal personnel, then it will be one huge horror story. And I can quite imagine that happening.

The negative attitude towards this tower block neighbourhood was especially widespread amongst our interviewees who did not have first-hand experience of it. People with more knowledge and especially those who have lived there or are still living there are very familiar with this negative image, but qualify it. They tend, for instance, to focus not on the social cohesion of this neighbourhood as a whole, but on the small-scale solidarity between immediate neighbours. They also value their personal freedom based on the lack of social control from a strong local community, which is still quite prevalent in Katwijk aan Zee. There is widespread agreement that there are more social problems and deprivation in this neighbourhood. There is, however, disagreement about the magnitude of these problems. These cognitive differences are magnified by the very different affective valuation of the difference between the neighbourhoods. There are also conative differences related to how local politics should or should not intervene to transform these deviant neighbourhoods into one which corresponds more with the identity of Katwijk aan Zee. For instance, every time the housing association carries out some maintenance on these tower blocks the local community in Katwijk aan Zee is awash with rumours that these much-hated tower blocks are to be torn down at last. They are seen as out of place in Katwijk aan Zee, where local identity discourses focus on the communal values linked to a traditional fishing community. The problems in this neighbourhood are also frequently debated in the municipal council and are linked to policies of neighbourhood renewal. Recent studies initiated by the municipality showed, however, that the problems in this and similar neighbourhoods were not only much less serious than assumed, but they also showed that the inhabitants of these neighbourhoods were quite content with their living conditions.

> In our opinion the living conditions were bad there. But they say: mind your own business. We live here fine. I close my front door and he closes his front door. He does not look at me and I don't look at him. No problem. It is a kind of people that did not like to care for each other. That kind of people loves living in that neighbourhood. There is drug abuse, syringes on the floor, rows and trash, you name it. But those kinds of people are comfortable with that. That does not fit the traditional Katwijkse identity, but ...
>
> (A local community leader)

Several interviewees who live in this neighbourhood further qualify the differences between the inhabitants of these flats and the community in Katwijk aan Zee. Many of its inhabitants are born and bred in Katwijk aan Zee, but have deliberately chosen to live in Katwijk Noord in order to be free of the group pressure and social control in the old fishing village. In this new housing estate, they have more individual freedom. This does not mean that it is a neighbourhood of isolated individuals. In this neighbourhood there are also social networks whose members support each other. But these are less visible and organised. They are also more focussed on direct neighbours and less on the local community as a whole. A manager of a local housing association recounts her discussions with inhabitants:

> People say: "I need fresh air, I want to do my own thing, in my own way. That is why I live here. I live first-rate. I know the whole area has an opinion, but it is my place." Thus there is a different kind of self-reliance, at a different level, but that fits that neighbourhood. It is a different way of living. Yes, it is more individual, but they check on each other. They really know what is going on in the neighbourhood and when necessary they help each other, but it is a conscious decision: "I need space to live my own life." And you don't have that many of such places in Katwijk. No, it is the norm, the rules, not washing on Sundays, all those kinds of stuff. And here is a place where you can live more anonymously, but this is immediately condemned from within the local community.

4.8 Anti-urbanism and local identity

Being different from urbanised spaces is a key element in how our interviewees discuss their local identity. They focus on "bad" neighbourhoods

not so much because of their objective housing and socio-economic conditions, but because of the different, more urban, way of life of the inhabitants of these neighbourhoods. Other fringe neighbourhoods with cheap
housing on the edge of the traditional villages are similarly negatively
stereotyped by established members of the local communities. One
member of a local neighbourhood council discusses the problems in a
neighbourhood close to the border with another village. He narrates a
cascade of problems, starting with dilapidated high-rise housing, inhabitants with problems, drug abuse, migrants, asylum seekers and ending with
single male Syrian refugees:

> That has nothing to do with being anti-foreigner. But they are all
> unaccompanied Syrians. I say I cannot give these people terraced
> housing when these come vacant. Or a three roomed apartment. We
> shall have to deal with that. That is of course a problem. We had a
> minor problem with Poles. In my opinion you should decently take
> care of them because these people work with us in fishing, in Rijns-
> burg in the flower trade, wherever. You have to deal with it. You
> cannot say come work with us and sleep in the street or in your car.
> That cannot be the case. We were working on that with the muni-
> cipality, the housing corporation and the tenant association. Writing
> nothing down, but exploring beforehand how we going to deal with
> that. Looking ahead. If next year 20 Syrians refugees arrive it will be
> war. I care about those things.

This opposition to the threat of external forces is also present in how our
interviewees characterise their own local community. Even though statisti-
cally its population density of 2,558 inhabitants per square kilometre
makes Katwijk officially an urban municipality (CBS 2016), they identify
themselves as a rural municipality. It is characterised by its inhabitants not
as a town of 65,000 inhabitants but as "villages welded together", or as "a
city which has remained a village", or as "the rustical centre of the Randstad". Some even flatly deny its urban character. "But also geographically
is Katwijk in my opinion no part of the Randstad, because I think about
the Randstad as an urban area. Living in Katwijk is so totally different
from living in Amsterdam."

Also on the island of Goeree-Overflakkee we found that the contrast
with the urban Dutch society was an important element in their local identity discourses. The urbanisation pressure appears to be marginal when
compared to Katwijk. Goeree-Overflakkee has a population density of just
185 inhabitants per square kilometre, which is just 7 per cent of the population density of Katwijk. In addition, unlike Katwijk, which borders a

large town, the nearest city to Goeree-Overflakkee is tens of kilometres away, across a sea arm with only two crossings. Whereas the dominant identity discourses in Katwijk opposes urban migration, the dominant regional identity discourse on Goeree-Overflakkee and its municipal policies focus on attracting new inhabitants from the nearby urbanised Rotterdam area (see also §3.1.1). Despite these huge differences in urbanisation and attitude towards urban migration, our interviewees on Goeree-Overflakkee also clearly distinguish their local community from the more urbanised Dutch society. They contrast their traditionally close-knit and rooted local community with the modern, dynamic, anonymous and mobile urban Dutch society as a whole.

> Look, we are very close to Rotterdam, but the mentality is completely different. Really totally different, incomparable. Here we work hard, we "just act normal, that's already crazy enough", don't get too big for your boots, and especially be modest, don't stand out from the crowd. You see, other entrepreneurs are very good in marketing their products. Yes, they behave differently than local entrepreneurs. They are aware of that. On the one side they admire it, on the other side they say: "just look at them". That is really culture, quite conservative. And of course there are here entrepreneurs who make the best products and sell them all over the world, but they don't show it in public.
>
> (Local administrator)

Modesty is a key characteristic of local identities on Goeree-Overflakkee and in most of the villages in Katwijk (with the exception of Rijnsburg). It is linked with their traditional village life and contrasted with an urban identity.

Local identities in general and the differences with cities in particular are regarded as threatened and declining.

> There is a tendency to guard against the outside world. Let us alone. On the other side of the island there is also a more traditional social structure; all sorts of things which have already disappeared a long time ago in the Randstad.
>
> (Local politician)

Many interviewees deplore the perceived loss of local identity. Besides some youths, all our interviewees value the non-urban character of their local identities. The city is, for them, the spectre they oppose. The threat of urbanisation has become a key component of their local identity discourses. They therefore tend to characterise the municipality of Katwijk with its high population density still as "a city which has remained a village". The threat of urbanisation and modernisation is thus an important part of their discourse on local identity. The next and final sections of this chapter further discuss the importance of this perceived loss of distinctiveness for local identity discourses. Whereas this section has focussed on the current differences with their urban neighbours, attention now shifts from these more spatial differences to the developments over time.

4.9 The loss of identity as a key element in identity discourses

Lagging behind in time compared to the rest of society is a common theme in our interviews. In reaction to the first very general question on the identity of Goeree-Overflakkee, a leading entrepreneur replied:

> Our rotation speed is lower than the rest of the Netherlands. I mean, we don't drive ourselves crazy by following hypes. We don't follow modernisation quickly. That has its origin in the island culture. Even though since 1963 it is no longer an island through bridges and dams. But the island is focussed on its own services, own people, own world, own culture. But life goes on in the rest of the world on these topics. But the culture is still here. It takes perhaps even four or five generations for the culture to slowly adapt itself. You see the difference between grandparents, parents, me and my children. My grandparents retain their culture, they cannot even speak standard Dutch, and my children who live in Amsterdam and Rotterdam, they still visit the island, but they can't even speak dialect. It also makes a lot of difference economically. In my perception it goes more slowly historically. And you will have heard that more often in these interviews, the just be modest principle, don't stand out from the crowd, that is important. Slowly but surely we have to be careful that we don't become a depressed area. Let's be less modest!

Time is an important element in how our interviewees discuss local identity. Their perception of time refers more to a general development path than to specific historical events and locations. The elements which characterise local identity are only loosely linked to history. History is only sometimes linked to the traditional local differences in agricultural specialisations, social structure, village life and religion, which are used to explain local identities and the differences between them. However, history in the sense of the lost good old days of the better past and the perceived decline in local community and identity is an important narrative which is widely used by our interviewees. Paradoxically, the feared loss of identity brings identity into the political debate. The current phase of liquid modernity undermines established identities, which ironically stimulates the importance attached to identity (Bauman 2004, 13–46).

Almost all our interviewees narrate about the loss of local identity. Many of them deplore this. The threat of urbanisation and modernisation is an important part of their local identity discourses. This loss has both spatial and temporal aspects. We already discussed the generally perceived threat posed by urbanisation to local identities. The strength and distinctiveness of local identities is also linked to the degree and speed with which specific local communities open up to the outside world. Their different experiences with urbanisation and migration are especially seen as explaining the differences in the strength of local identities. This was the case not only for the urban neighbourhoods, like Katwijk Noord, discussed in §4.7. Our interviewees also make more subtle distinctions between the different local communities in their municipality. On Goeree-Overflakkee, its western part is generally considered as more traditional, while the eastern part is regarded as more like the rest of the Netherlands. In the municipality of Katwijk, the fishing village of Katwijk aan Zee is regarded as the most traditional community and the small village of Valkenburg is generally seen as the most closed community. This sets them apart from the more urban and indistinct sub-urban villages of Katwijk aan de Rijn and Rijnsburg.

This exemplifies a narrative of authentic and traditional places being gradually flooded by urban influences, radiating from the cities and leading to the successive reduction of the individual identities of places. This assumed diffusion transforms distance into a time lag. This is quite a common conceptualisation. The most rural peripheral areas are generally considered as having the strongest identities. The romantic conceptualisation of the unspoiled native living far away from the depraving influences of urban civilisation is still a powerful image. This results sometimes in a residual conceptualisation of local identities. Local identities are often conceived as relics of the distant past in the far away periphery, which will disappear over time and which can thus be largely ignored in politics and administrative reforms.

This loss of local identity due to the homogenising influences of modernisation in general and urbanisation in particular is also an established topic in academic circles. For instance, a study published in 1949 of a local community in the Dutch periphery also concluded that there was a decline in local identity and linked this to modernisation and urbanisation (Groenman & Schreuder 1949). The founding fathers of the social sciences discussed in the nineteenth century the decline of local identities and also linked it to modernisation and urbanisation (Bryant & Peck 2007).

Local and other identities always position themselves towards the past and the future. Visions of both the immediate and the distant pasts and futures are part of identity discourses. The intergenerational continuity of communities is an important aspect of their identity. Anthony Giddens (1991) stresses the importance for identities of their reflexive awareness and discursive consciousness for dealing with existential anxiety. Their narratives therefore must be kept going by incorporating new threats into identity discourses. Uriel Abulof (2015) studied how small threatened nations (Quebecois, Afrikaner and Jews) use their identity. He found that the idea that the current generation was the last generation with a distinct identity was an important part of their current identity discourses and political actions. The ideas of being the last generation and the feared mortality of the collective identity are crucial for understanding these identity discourses and how they are used in politics (Abulof 2015).

This is similar to the conceptualisation of many of our interviewees of their local identity. The younger generation is seen as not interested in issues of local community and identity (see also §4.3.3). Or in the words of a community leader: "In the past of course the older generation had a genuine identity and were very close, but with all those inhabitants coming from all different places it is no longer like that." We can only speculate what explains this. Is it just that younger people are, because of the challenges in their stage of life, temporarily locally indifferent (§4.3.3) and will over time become more attached to the local community, when they put down roots through buying a house and raising children? Or will specific local identities disappear over time as already predicted a long time ago? Or is this perceived loss of identity an element in all identity discourses? Primordialists discern, for instance, in the past the true authentic roots of their nation which have to be protected against the constant dilution by migration and modernisation (Storey 2012). But perhaps differences in local identities are not so much disappearing as being transformed. Modernisation and urbanisation takes many different forms and affects different places in different ways. This will, on the one hand, undermine existing local identities, but on the other hand create new differences. Traditional local identities thus constantly diminish, while new and different local identities are simultaneously formed. In the selective collective memory, this is seen by many as a loss, but by some as a gain, creating new local identities.

> What has increased, certainly in the last 40 years, is the involvement with the city. Not only the material connections have improved over the years, but also the city comes this way. People settle here, live here, who do not come from here. And also holiday-makers. I think

that we in the meantime, very sneaky quite substantially have changed, in our DNA. Thus the urban influence has changed our identity very much. Not that we are suddenly very different, but you notice: something is added to it.

(Local businessman)

4.10 Traditionalist and modernist identity discourses

Local identity discourses not only express the characteristics of the local community, but also suggest a developmental path. They formulate narratives on how identities were formed in the past, how these still partly exist in the present and how these will change in the future. The dominant narrative is of clear differences in the past between local communities, which now have become blurred through processes of social change, like urbanisation and increased mobility. The different local identities are still considered as important, but the diversity is expected to further diminish or even disappear in the future. Our interviewees broadly share this narrative of still important but declining local identities. But as has already been discussed in this chapter, our interviewees sometimes value these local identities very differently. Many fear this uniformisation of local identities and oppose it. Others welcome it and want to promote this development.

Figure 4.5 depicts these different perspectives of traditionalists and modernists towards changing local identities. The experience of the present generation is linked to their visions of the distant past and future. There is widespread agreement among our interviewees that local identities were stronger in the past and will further weaken in the future. This cognitive agreed weakening of local identities is valued differently. This divergence is simplified in Figure 4.5 in the opposition between traditionalists and modernists. They diverge not only affectively but also conatively; they value this development very differently and as a result support different political actions. The traditionalists want to preserve and defend as much of their traditional thick local identity, while the modernists want to open it up and link it more with other identities at other spatial scales. This difference is rooted in a different view of the distant past. The traditionalists value the strong cohesion and collective identity in traditional villages which formed tightly knit and safe communities and which were distinct from each other. Modernists agree on the tightly knit character of traditional villages, but value this differently. They object to the lack of freedom for the individual to choose their own way of life and determine their own future. The strong bonding in traditional villages is seen as smothering individual freedom. The modernists see the decline of the traditional village as part of the emancipation of the oppressed individual and

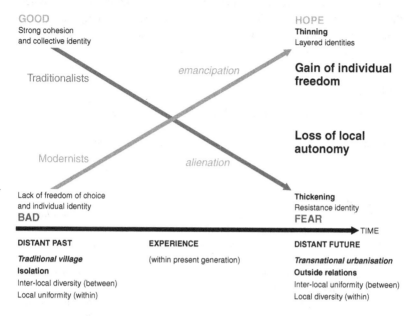

Figure 4.5 Traditionalist and modernist identity discourses of fear and hope.

the emergence of a more open, modern society. The traditionalists abhor the decline of traditional villages and local communities with a distinct local identity and fear its disappearance. Their fear can lead to thickening identity discourses and the formation of resistance identities (§2.8). Modernists, on the other hand, welcome the further decline of traditional villages and thick local identities. Their local identity is linked to many other layers of spatial identities in this globalising and urbanising world (§2.7).

This difference is very similar to a frequently made distinction between two dominant types of nationalism: civic and ethnic nationalism. Civic nationalism focusses on modern rights and freedoms, while ethnic nationalism stresses the importance of descent, heritage and history (Kaufmann 2008).

In local communities, there is widespread agreement on the cognitive content of local identities. Despite this cognitive agreement, different groups can strongly disagree on how they value these identities. In the face of these affective discords, local communities have however learned over time to cope with these differences in their everyday life. Municipalities with different local communities tend to develop over time their own method of dealing with these different visions on identity within a local community and the different identities between local communities. This becomes a kind of secondary identity, as was discussed in §2.9. When these secondary identities disappear during a municipal amalgamation, the feared threat to primary local identities can become a potent force in local politics. This is discussed in the next chapter.

References

Abulof, U. (2015). *The mortality and morality of nations.* New York: Cambridge University Press.

Bauman, Z. (2004). *Identity: conversations with Benedetto Vecchi.* Cambridge: Polity.

Bryant, C.D & D.L. Peck (Eds) (2007). *21st century sociology: a reference handbook.* Thousand Oaks: SAGE.

CBS (2016). *Gemeente op maat: Katwijk.* Den Haag: CBS.

Elias, N. & J.L. Scotson (1965). *The established and the outsiders: a sociological enquiry into community problems.* London: Frank Cass.

Giddens, A. (1991). *Modernity and self-identity: self and society in the late modern age.* Cambridge: Polity.

Groenman, S. & H. Schreuder (1949). *Sociografieën van plattelandsgemeenten: Ommen.* Den Haag: Staatsdrukkerij.

Hirschman, A.O. (1970). *Exit, voice, and loyalty: responses to decline in firms, organizations, and states.* Cambridge, MA: Harvard University Press.

Hogenstijn, M. & D.P. van Middelkoop, (2008). *"Zo werkt dat hier niet": gevestigden en buitenstaanders in nieuwe sociale en ruimtelijke kaders.* Delft: Eburon.

Hogenstijn, M., D.P. van Middelkoop & K. Terlouw (2008). The established, the outsiders and scale strategies: studying local power conflicts. *The Sociological Review,* 56, 144–161.

Hummon, D. (1992). Community attachment: local sentiment and sense of place. In: I. Altman & S.M. Low (Eds) *Place attachment.* New York: Plenum.

Kaufmann, E. (2008). The lenses of nationhood: an optical model of identity. *Nations and Nationalism,* 14, 449–477.

Lewicka, M. (2011a). On the varieties of people's relationships with places: Hummon's typology revisited. *Environment and Behavior,* 43, 676–709.

Lewicka, M. (2011b). Place attachment: how far have we come in the last 40 years? *Journal of Environmental Psychology,* 31, 207–230.

Storey, D. (2012). *Territories: the claiming of space* (second edition). New York: Routledge.

5 Municipal amalgamation and changing local identities on Goeree-Overflakkee

Municipal amalgamation is one of the most drastic and visible changes local politics has to deal with. This was the case in both Katwijk and Goeree-Overflakkee. This chapter focusses on the origins and aftermath of the use of local identities in the merger of the four municipalities on Goeree-Overflakkee in 2013. The next chapter analyses the use of local and regional identities in the local politics of Katwijk, which was amalgamated in 2006. These two chapters show that over time and between places local identities can be used in local politics in very different ways. These differences will be further analysed in the concluding chapter.

This chapter starts (§5.1) with analysing the use of local identity by the opponents in Goedereede against the amalgamation of their municipality. The decline of this local resistance identity discourse is examined in §5.2. Then attention shifts in §5.3 to the dominant role of local entrepreneurs in the formulation of a thin regional identity discourse focussing on the development of the island as a whole. We analyse how this thinner regional identity discourse was constructed by the business sector on the island and how it became the dominant identity discourse through its alignment with elements of existing identity and policy discourses operating at different spatial levels (§5.4). After amalgamation, spatial identities were not only used to connect local communities, but also became important elements in some political conflicts in the municipality. The aligning of identity discourses focussing on economic development coincides with the growing apart of cultural identity discourses focussing on religious and individual freedom, especially in relation to the regulation of Sunday rest. This is discussed in the last section (§5.5) of this chapter.

5.1 An emerging local resistance identity before amalgamation

After several unsuccessful attempts to strengthen the cooperation between the four municipalities on the island of Goeree-Overflakkee, the Dutch central government decided in 2011 that an amalgamation was necessary to create an effective local administration on the island. This met with stiff resistance from the administration and population in the municipality of Goedereede. In its official petition against amalgamation to the province of Zuid-Holland, the municipality of Goedereede formulated as its first argument that its local identity differed from the rest of the island. "There is no recognition of the separate position of Goedereede and its inhabitants as a distinct part of the island" (Goedereede 2010). Their other arguments focussed on its administrative strength and the lack of urgency to amalgamate all the municipalities on the island. The resistance against amalgamation was also very strong among the population of Goedereede. In May 2009, 89 per cent of those polled (4,825 out of a total population of 11,375) were opposed to amalgamation. But despite the opposition of the local population, the municipal council and the local administration, the Dutch government pressed ahead with the amalgamation and on 1 January 2013 created the new municipality of Goeree-Overflakkee. A few months after the amalgamation, the vast majority of the population was still against amalgamation. The elderly especially (88 per cent) were against the amalgamation. The supporters of amalgamation were, in contrast, predominantly young and not born in the municipality. A survey of the population on the reasons why people opposed amalgamation clearly showed that the feared loss of local identity was the dominant motive. A possible increase in taxation and dissatisfaction with the imposition of amalgamation by the government were far less important reasons for opponents to resist amalgamation (Jeekel *et al.* 2013, 30–56).

The perceived necessity to protect the local identities against new developments outside the control of the local communities fuelled the opposition to amalgamation. In the old municipality of Goedereede, there was wide opposition against amalgamation in the villages of Ouddorp and Goedereede. In Stellendam, the third village in this old municipality, a few kilometres to the east and with a very distinct history, there was much less opposition to amalgamation.

The strong resistance in these two local communities changed the character of their local identity discourses. They became more inward oriented, focussed more on their historical roots and their differences with others; they "thickened" into a resistance identity (§2.8). This resistance identity discourse on the uniqueness of the old municipality of Goedereede was clearly present in the documents we studied before our interviews. We therefore confronted our interviewees with a picture of one of the placards used by the opponents of amalgamation. This placard promoted the re-flooding of a polder in order to re-establish the situation before 1751 when Goeree was still a separate island. A migrant from outside the island recognised this instantly.

> Of course, in the past it was two separate islands. They were separated by mud and water. Thus they used to be separate. And that is not a long time ago. There are still people alive who can remember that those two were separate.

He continued to stress that these divisions are still noticeable today and that people from both parts have few relations with each other compared to the relations they have in the western part of Goeree-Overflakkee with neighbouring coastal regions and in the eastern part of Goeree-Overflakkee with the more urbanised inland areas. All our interviewees acknowledge the differences in identity on the island and most of them link the resistance in the municipality of Goedereede with the feared loss of autonomy and the anxiety among the local population that this would undermine their specific local identity. A leading supporter of municipal amalgamation elaborates on these arguments against amalgamation.

> We used to be two islands with water in between. There are also differences in dialect between the western part and the rest. That is also linked with different views. They had quite a prospering municipality Goedereede and they successfully managed their own affairs. That has to do with the identity of we can manage our own affairs. And now we have to be a single municipality? They fear this will water down their identity, also politically.

This broadly fits the discussion of the emergence of resistance identities in §2.8 and is quite similar to the analyses of the opposition against the amalgamation of Nurmo discussed in Box 2.5.

5.2 The fading local resistance identity after amalgamation

In our interviews, which were conducted some years after the amalgamation, this opposition to amalgamation was frequently re-evaluated and downplayed. Although our interviewees acknowledge the differences in local identities on the island and the role they played in the resistance against amalgamation, almost all our interviewees reacted with embarrassment to the picture of one of the placards used by the opponents of amalgamation which suggested the re-flooding of a part of Goeree-Overflakkee to divide it into two separate islands. "I think this is just very emotional, against their better judgement. And really when I now look at it I am ashamed ... such emotional utterances!" (local resident). This passionate focus on differences in local identities is now regarded as incompatible with the local identity, which is also based on the respectful dealing with differences.

> The resistance here was very strong. But there is also an attitude of going for it. Now we are one and we must bury the hatchet and forget it ... it is how things are, yes we are now one. And yes, that has something to do with identity. When the amalgamation was there, everyone wanted to share in the spoils and try to make the best of it. I appreciate that, and it makes us special.
>
> (Employee in the tourist sector)

When asked about the legacy of the resistance to amalgamation, a local politician and opponent of amalgamation comments:

> It has gone silent. People who know I opposed it frequently tease me and ask me: "what happened to that opposition?" But people resume their everyday life, and for most of them nothing much has changed. Initially people were angry that they had to travel all the way to that new huge town hall. But that is peanuts, either you go by car to Middelharnis, do it over the internet or chose home delivery, that's not what it is about. It was more about culture, feelings, identity. This brings us again to the issue of identity, which involves much more than religion. It focussed on which village you live. That was what it was about. When do you need a new driving license? Once every ten years. You don't make a fuss about that! But the brutality of that amalgamation, which was forced down your throat, and knowing that you have been taken for a ride and that the arguments of the government are rubbish, that has disappeared. The drawing of that line wipes that out.

This suggests that the use of identity in the resistance against amalgamation is a temporary phenomenon which disappears after amalgamation, although some resentments live on and the gap between administration and population widens.

Even the leading opponents of the amalgamation we interviewed were now more or less ashamed of the emotional resistance they had organised.

Contrary to the documents they produced only a few years ago in their attempts to oppose amalgamation, they now tend to deny or at least downplay the role local identities played in the protest. They now state that the opposition was more based on general arguments, like the assumed benefits of amalgamation through increased efficiency and effectiveness. They no longer stress the incompatibility of their local identity with the identities of the local communities in the rest of Goeree-Overflakkee. A local politician and former opponent of amalgamation living in Ouddorp told us:

> People here were against amalgamation. In a poll at the time 90 per cent were against, but now the amalgamation is completed, they are also law-abiding. That typifies the local identity. People loyally contribute to the formation of the new municipality. It is no use to look backwards, like it was, we never get that back. Then better make the best of it and exploit the new possibilities.

This and other organisers of the resistance against the amalgamation are now remarkably active in public bodies within the new municipality and in initiatives to promote the development of the island. Now their municipality of Goedereede no longer exists, they try to use the new municipality of Goeree-Overflakkee to promote their local interests. They do not like to look back in anger, but want to look forward in the expectation that the new municipality will be instrumental in the promotion of a new, attractive island identity, which will support their local interests and identity. The formation of the new municipality created a new positive and forward-looking dynamic, which makes them look back in embarrassment to the negative emotions before the amalgamation. They seem embarrassed by the old identity discourse of their threatened local identity. The focus in their identity discourse has reversed. Their backward-looking, thickening local identity discourse resisting amalgamation has more or less disappeared with their old municipality. Now they promote a forward-looking, thinner regional identity, which is rooted in their local identity and links up with similar local identities in all the villages on the island. Their dominant strategy to protect and promote local identity and interests has shifted from the preservation of their old municipality to the development of their island through the presentation of an attractive island identity to the outside world. This regional identity discourse was formulated years before the amalgamation by local entrepreneurs. How this came about is discussed in the following sections.

5.3 A thin island identity formulated by local entrepreneurs

In 2006, several years before the municipal amalgamation in 2013, the three local branches on the island of the cooperative Rabobank merged. This is the main bank financing rural entrepreneurs. This merger process stimulated the formulation of a thinner regional identity discourse. The management of this bank actively promoted municipal amalgamation and in 2006 also produced a vision document analysing the island's economic problems and possible solutions. Their vision of the future of the island focusses on reversing the decline of the peripheral island economy through economic restructuring, attracting more inhabitants and making better use of the touristic and residential qualities of the island. This must solve its worsening peripheral position and transform it into a sustainable island. Box 5.1 elaborates on this vision of the business community, which is based not only on this and a later document produced by the Rabobank (Rabobank 2006, 2014), but also on a document of the cooperation of local business associations on the island which presents a very similar vision of the future of the island (IPGO 2011).

Box 5.1 The vison of the business community: towards a sustainable island economy

The vision of the business community is formulated in three key documents produced through the cooperation between key members from different business associations from the whole of the island and the cooperative Rabobank (IPGO 2011; Rabobank 2006, 2014). These documents present a similar perspective on the economic characteristics, challenges and policy choices for the island. These documents are frequently mentioned in the interviews and the views expressed in these interviews correspond with this vision.

The characteristics of being an island dominate the spatial conceptualisation of the business community. The island is presented as a physical object and as an undisputed territorial unit. The economic structure and problems are analysed as located on the island and mostly presented as characteristics of the island as a whole. The island is also the object of political action and economic policies. Great importance is given to the cooperation between the amalgamated municipality and all the different sections of the business sector on the island, in order to further the common interests of everybody on the island. They are convinced of the necessity of cooperation to initiate new policies in order to restructure the island economy, to get support from the outside world, by accessing new urban markets and attracting subsidies from higher level administrations.

The problems of the island are linked to the peripheral position of the island. Goeree-Overflakkee lags behind other Dutch regions on many comparative economic indicators of performance like growth and economic structure. Its economy is dominated by traditional sectors with low-skilled jobs and a relatively underdeveloped service sector. The island suffers from the brain drain of skilled youth to the urbanised regions of the Netherlands. The island is portrayed as in danger of even further peripheralisation.

This negative spiral has to be reversed. The island economy has to become more innovative and sustainable, with a larger and more modern service sector, so that the business sector and the population in general can profit from the possibilities of the new economy. This will reverse the feared wave of out-migration from the island. The island economy should achieve this by profiting more from the competitive advantages associated with the characteristics of the whole island. Its residential qualities should be used to attract more and wealthier inhabitants. Some of the attractive residential characteristics are based on the situation of the island, such as its close proximity to the urbanised Rotterdam area. Others are based on the existing site characteristics of the island, such as the coast and the rural landscape. All these existing characteristics have to be further developed and augmented with new specialised services to cater for the needs of the nearby urban population.

The comparative advantage of the touristic sector also has to be restored. Tourism has flourished since the island became accessible in the 1960s through an improved infrastructure based on dams and bridges. But because it is based on a diminishing number of campsites near the beach, it needs to be upgraded. New, more upmarket leisure and watersports facilities along the coasts of the whole island have to be realised – for instance, through redeveloping the dozens of old harbours spread all over the coastline of the island. More people must be attracted from the nearby urbanised Rotterdam area, not only as day tourists, but also as new inhabitants, in order to widen the population base for essential public services like hospitals and local shops. Besides the building of attractive housing, this also requires the preservation and further development of an attractive coastal landscape and an improved service sector focussing more on health and wellness. This will also provide employment opportunities for the well-educated youth, enabling them to remain on the island.

This has to be done in a sustainable manner with a focus on ecological improvements. These are linked to creating a more attractive rural landscape for inhabitants of the nearby Rotterdam urban region. This relatively empty and windswept island with strong tidal currents can also become an energy self-sufficient and sustainable island, using new technologies to create additional high-skilled employment. Active island marketing is necessary, in cooperation with the new municipality, to promote the island and communicate its forward-looking identity, not only to the nearby urban markets, but also to the provincial and national administrations, in order to attract the necessary support to implement the above discussed policies (IPGO 2011; Rabobank 2006, 2014).

5.4 The unification of the local entrepreneurs

> We need some big businessmen to lay down the framework and the
> rest will follow.
>
> (Prominent local politician and municipal administrator)

Concerns about the economic weakness of this island are not new and have, for many decades, been the subject of central government policies. The central government has invested heavily in reconstruction, flood control and infrastructure, especially in the aftermath of the flooding in 1953, which inundated three quarters of the island. In the 1960s, the improved connections with bridges and dams gave entrepreneurs on Goeree-Overflakkee new opportunities, especially in the tourist sector. But this economic sector is becoming obsolete, or as a local administrator puts it:

> It used to be farmers combining their farm with a site for 10 or 20
> caravans. They yearly returned there for 20, 30 or 40 years on end.
> Now they are too old for it and young people are here today but gone
> tomorrow. Local entrepreneurs adjust too slowly to that trend.

A prominent touristic entrepreneur remarks that: "a large section of the population is convinced that something has to change to break the lethargy". But help from central government can no longer be taken for granted in this era of globalisation and neoliberal restructuring of state policies. Instead, many entrepreneurs became convinced of the necessity of strengthening their cooperation on the island. A leader of the federation of business associations for the whole island, established in 2013 – who is also a former leader of the opposition against amalgamation – comments:

> People say we are one island and if we don't do it, nobody is going to
> take care of us. That is based on the mentality of both the western and
> eastern part of the island. And now you see this also in the business
> community; yes we are responsible and we are going to tackle it. We
> are going to cooperate and act. Because there is nobody else. The
> province ignores us. We have to do it ourselves.

Business interest were traditionally organised very locally on Goeree-Overflakkee. This allowed, for instance, the touristic entrepreneurs on the western coast of the island to have direct access to their own municipality. Individual entrepreneurs could also easily play municipalities against each other to obtain the best and cheapest new business sites. However, as the economic problems became more apparent at the start of this century, this political fragmentation was increasingly seen as a hindrance for the promotion of business interests. A prominent member of the housing and building industry comments on the prominent role of the business sector in the amalgamation of the municipalities on the island:

> The business world wanted to deal with only one administration here on the island, for practical reasons, for the economy of the island. Goeree-Overflakkee is a geographically well-defined region where everything is intertwined, but where until recently four municipalities pursued their own policies and each separately wanted to please their inhabitants. Thus everything on the island was four fold. Four business parks.... That is of course undesirable, because you have to consider the interests of the whole island. You must reason what is the best location, not only for businesses, but also for other facilities, like education, health care, recreation. We must not try to score points off each other, that creates the mutual animosity of wanting to outperform each other. Then you put four times energy in the same. On an island with only a population of 48,000 this is really absurd. That's why we said, we want in the future a single municipality, who together with others, will solve the problems like a social entrepreneur.

The amalgamation process of the four municipalities on the island coincided and was intertwined with the growing cooperation between local business associations and the development of a shared regional identity communicated through island marketing. A leading figure in the business community gives his view on this interconnection between the political and economic forces on the island.

> It reinforced each other. My business association was the largest and had members all over the island. At a certain point we agreed that we are absolutely in favour of amalgamation because we as entrepreneurs want a substantial partner and uniform policies. And that we as entrepreneurs should also become more united. Then my business association, together with that of the western part of the island joined forces, creating an overarching association, not for social events – which is the focus of most business associations – but to promote business interests, not individually, but for entrepreneurs of the whole island. We took advantage of the development towards municipal amalgamation. If we as entrepreneurs want an amalgamated municipality, let us as entrepreneurs form an association uniting all entrepreneurs on the island. It helped that we could say that we lagged behind the process of municipal amalgamation, whose sluggishness we always criticised.

Thus every business association appointed a board member for the new federation. A flow has been created between the municipality and this federation. The municipality has ambitions, the entrepreneurs have ambitions; present the attractiveness of the island, quicker decision making, closer social and political relations, being more proud of the island. A momentum has been created supported especially by entrepreneurs. The daily life of the average citizen is however not much affected by amalgamation, they still shop in their own village and want to keep things as they are.

Local entrepreneurs are very convinced and proud of their leading role in the amalgamation process and the formulation of an economic vision on which the new regional identity discourse of the island is based. A touristic entrepreneur comments:

I think the solidarity will increase, I believe that it is already increasing. I like that. Because, despite that I experience it as an entrepreneur, entrepreneurs are also citizens, thus they pass that message on. And entrepreneurs oversee employees, people from the island. More and more interaction is generated and that is good.

A prominent local entrepreneur comments on the leading role of the local entrepreneurs:

It slowly trickles down. When you lead the way then you sometimes look back and in the rear people are unaware of who leads the way. Figuratively speaking that is. But if you communicate plentifully in many different ways, in the local media, organise meetings, municipal bulletins, then it reappears (...). A kind of brainwash is necessary. But we are in only at the beginning of the Gaussian curve to adjust the identity. This is evolution not revolution. That does not suit Goeree-Overflakkee. You have to do that slowly, in small steps.

5.4.1 Island marketing

The focus of the entrepreneurs on the island economy as a whole does not mean that they promote their interests only using the amalgamated municipality. Other levels are also used to collectively promote the economic interests of the local entrepreneurs on the island to the outside world. A prominent member of the housing and building industry stresses the importance of the relation between amalgamation and economic interest articulation to other administrative levels.

We speak ever more with a single voice to higher authorities like the province. But you not only have to deal with the province, but also with the national administration and cooperate in projects on, for instance, water and nature. There we don't speak with one voice yet. If we discuss employment and the economy then we consider the whole island. In my opinion, but I have always supported it, in my view, everyday a new element falls into place. Gradually politics gets more control and progresses. We are moving in the right direction.

In the words of a local bank manager: "Only the last three years the province takes notice of and confers with Goeree-Overflakkee. In my opinion while there is now one municipality; that amalgamation is so crucial!"

Island marketing helped the promotion of the economic interests of the island towards higher levels of administration. It became an important topic in the cooperation between the business federation and the municipality and shows how initiatives of the business communities on the island influenced and, to a large extent, shaped the economic policies of the new municipality. Island marketing was also based on some older local initiatives.

In Ouddorp a small group was promoting their products. Also in Middelharnis there were four or five small groups with their own ideas about marketing. This became linked to the process of amalgamation and the creation of a federation of business associations on the island.

This local administrator gives more details of the leading role of entrepreneurs.

Three entrepreneurs, supported by business associations, saw the opportunities of the amalgamation to conceptualise the island as one. They raised funds from the chamber of commerce, the Rabobank and some business associations. These three organised a three day trip to the island Texel with three spearheads. Island marketing; how did they create the image of that island, and how can we do that here? Another spearhead was business associations. We have 15, on Texel they merged into 2. How did they do that? The idea was that business associations should unite to take a stand, to speak with one voice to the amalgamated municipality. The federation of business associations was formed to achieve this. Regional products was the third theme. How do they do that on Texel and how could Goeree-Overflakkee do it? And you see that all three are put into practice.

The marketing of regional, mostly agricultural, products stimulates cooperation between different business sectors on the island and the sense of living on the same island by the whole population. This communicates the new regional island identity, not only to the outside world, but also to the producers and the inhabitants on the island. (See also the example discussed in Box 5.2.) A factory owner links traditional elements of the identities of local communities with the value of the regional products, and uses this to sell it to the outside world.

> I supply my cheese to the Albert Hein supermarket and my ice to Lidl. Albert Hein through their office in Culemborg and Lidl in Huizen; way outside this island. The location is less important than the story. I don't sell cheese or ice, I sell a story. If you want to do business with Lidl, your story has to be excellent. Price is important, but the story is crucial. I thus focus on regional products, but as part of a wider story. We are organic, are regionally produced, a family firm, everything produced on the island, even on a care farm. This makes a nice story.

These initiatives to promote an attractive regional identity to the outside world have to deal with existing local identity discourses. As corroborated by our interviewees, elements like close family ties, hard work ethic, resilience, self-reliance, inventiveness, solidarity and sense of community are important in all the different local identities on the island. These traditional thick local identities are combined with the attractive Dutch open landscape, the beaches of the island and their proximity to Rotterdam. These established elements of local and regional identities are also linked to new policies for sustainable regional development. All these old and new elements are aligned in a new, thinner regional identity discourse which is used to promote the island and to mobilise support for the new municipality from the population.

5.4.2 From local to regional identity discourse

On Goeree-Overflakkee, the local entrepreneurs united to collectively articulate their interests to the newly amalgamated municipality on the island. This cooperation is based on a growing awareness that the lagging island economy can only be improved when all stakeholders act together. The business community took the lead in formulating the economic policies and priorities of the amalgamated municipality and the promotion and communication of a new regional identity through island marketing. They were leading in the conceptualisation of the island as a territorial unit where everybody shares the interest of counteracting the relative economic

Box 5.2 Cooperation to market local products: *Buutegeweun*

Figure 5.1 The stand of *Buutegeweun* in the new covered market in Rotterdam.
Source: photo taken by Maarten Hogenstijn.

Entrepreneurs cooperate in different organisations on the island of Goeree-Overflakkee. The FOGO, the federation of local associations of entrepreneurs, focusses on the municipal and provincial policies. The cooperative society *Buutegeweun* brings together farmers, fishers, breeders, bakers, butchers and horticulturalists. Together they promote the sale of regional agricultural products. They use the trademark "*Buutegeweun*", which translates as "extraordinary". This brand name is a playful transformation of "just act normal", which is also frequently mentioned by our interviewees as an important characteristic of the identity of the local communities on the island. This also sets the rural islanders apart from the expressive and boasting city-slickers of nearby Rotterdam. Under the brand *Buutegeweun*, they sell in Rotterdam many different traditional, local agricultural products from all over the island, like fish from Stellendam and potatoes from Flakkee. The attractive, unspoiled rural landscape of the island and its partially organic regional products are linked to sustainable production in family-based farms that also employ people with disabilities. This brand is thus rooted in thick local identity elements based on traditional, local agricultural specialisation and community values of solidarity and self-reliance. It conflicts with some other thick identity elements. They try to overturn the traditional modesty over their own achievements and qualities by using the brand *Buutegeweun*, which boasts the extraordinary quality of their regional products. These were, until its closure, in December 2014 sold in a stand in the prestigious new covered market in downtown Rotterdam. Their opening hours on Sunday, however, clash with the Christian identity of many communities on the island. One of the involved businessmen remarks that: "What you have initiated for the whole island and have designed to be inclusive for everybody is obstructed by an orthodox religious minority. You can't do anything against it." When asked about changes in local and regional identities on the island, he continues:

> With unification things have changed obviously. Initial resistance was very visible, but that has changed. There is now perhaps more identification with the island than before. Everybody does one's thing. The municipality tries, and I gladly contribute, to make it more visible to the rest of the world. That has changed a bit I think. And that generates some pride.

decline, caused by an outmoded uncompetitive economic structure. They want to use the attractive characteristics of the landscape and the communities of the island to promote a new and appealing regional identity to the outside world. The entrepreneurs successfully conceptualised that becoming more attractive to the outside world is the key for economic development. Communicating this new regional identity must attract people from the nearby urbanised Rotterdam area and help to obtain resources from higher levels of administration.

Local communities on the western tip of Goeree-Overflakke perceived amalgamation as a threat to their identity. Opponents emphasised their distinctiveness and historical roots in their identity discourse. This thickened resistance identity was successful in mobilising the population to oppose amalgamation, but could not prevent the amalgamation imposed by the government. After the amalgamation, local stakeholders could no longer protect the local identity and interests through their own municipality. Instead, some of them joined forces with entrepreneurs who had been advocating the municipal unification of the island to promote their local interests and identity. In their strategy, they jumped scale from the local (old municipality) to the regional (new municipality) and now try to protect their local identity and interests through the development of a new, thinner regional identity discourse, which is still rooted in their local identities.

5.5 Sunday rest: the accommodation of different local identities in an amalgamated municipality

Many in the old municipality of Goedereede perceived amalgamation as a threat to their identity and autonomy. They felt that this would put an end to their established way of dealing with local identity issues and that the locally negotiated ways to accommodate the different traditionalist and modernist views on local identity would be threatened (§4.10). These secondary identities (§2.9) are mostly quite subtle and informal. This section discusses in more detail how local politics deals with the different views and interests towards Sunday rest, which is one of the contested topics in municipal politics on Goeree-Overflakkee.

In municipalities, over time, there emerge informal ways to manage the different views on this and other topics related to local identity. The main difference, as discussed in §4.10, is between the modernist attitudes of reducing the limitations posed by local identities to new developments and the traditionalist attitude of protecting existing local identities. Dealing with these differences creates a kind of secondary identity of how to deal with different local identities and different perspectives about primary local identities (see §2.9). Through amalgamation, this secondary identity disappears with the political arena of the old municipality. This catapults the protection of, and disputes about, the more primary and well-established local identities into the political debate. The emergence of thickening resistance identities before amalgamation is one consequence. Another consequence is the emergence of political conflicts in the new municipality – for instance, in dealing with different views on Sunday rest in the amalgamated municipality. This example of the use of local identities in local politics in an amalgamated municipality is analysed in this section.

We did not ask our interviewees explicitly about the role of Sunday rest in relation to local identities and politics. But different opinions about Sunday rest were discussed by some of our interviewees in Katwijk and by many of them on Goeree-Overflakkee. Respect is the key word used by many of our interviewees in their discussion of Sunday rest; respect not only for the observance of Sunday rest, but also for those who have a different opinion on the role of religion in local politics, especially in relation to the regulation of Sunday rest. Respect was mentioned 54 times in all interviews, 43 times on Goeree-Overflakkee and 11 times in Katwijk. On Goeree-Overflakkee, this was, with one exception, always linked to religiosity in general and Sunday rest in particular. In Katwijk, most of the time it was linked to respect between the four different local communities.

Respect is generally seen as an important characteristic of cohesive local communities. It is very typical for secondary identity discourses dealing with different primary identities, as discussed in §2.9. It is a concept used in most local identity discourses. Or in the words of a local administrator: "You all live together and of course we are all different, but you just respect each other. Yes, I like that." This norm of respect is discussed in relation to the daily life of individuals, the relations between different local groups and the behaviour of political parties in local politics. Many interviewees, especially on Goeree-Overflakkee, discuss this in relation to how, in their daily life, individuals comply or deviate from the social norms of their local community. Respect for Christian religiosity in general and Sunday rest in particular is central in these discussions. Many also link this to a perceived lack of respect of the dominant religious groups and parties for non-religious people. A local politician from the western part of Goeree-Overflakkee expresses this as follows:

> I behave myself. I won't paint my house on a Sunday. I take that into consideration. But they must not swing that around. They must not tell me what I can do and cannot do on a Sunday. That is rubbish. That really annoys me.

A born-and-bred local businessman who personally does not observe Sunday rest comments on how he deals with the issue.

> It is quite intricate based on individual experience, but it really does not bother me at all. If they are closed Sundays fine, I go elsewhere. It is also quite nice to have some peace and quiet, you know. I don't mow my lawn on Sundays, for instance. I adjust myself to the environment.

This is a very common view among those who themselves do not observe Sunday rest. Some claim that the observance of Sunday rest even attracts non-religious people to Goeree-Overflakkee. A leading local businessman in the housing sector comments:

> There are now also groups of migrants in Ouddorp who vote in national elections for the liberal party (VVD) but vote in the local elections for the party of orthodox-protestants (SGP), just because they want to preserve tranquillity, as that's why they moved to the island.

Not all migrants are actively attached in this way to the local community (see also §4.3). Others are more indifferent to the dominant local identity and, as a result, experience difficulties in becoming accepted by the local community. Their behaviour based on their local indifference in a way further strengthens their local indifference. According to the same local businessman:

> You have to respect the dominant culture. There are of course those who do not like this. They just say: "I am accountable to nobody and

just do what I find right." These position themselves at the margins of society. And then they complain that they are never accepted by the local community, however long you live here. But that has to do with the fact that their way of life differs from the dominant culture. That is in itself not problematic, but it is when you explicitly disregard others. That is the issue. If you live in a village where Sunday's church attendance is still high, and you deliberately mow your lawn when people go to church or wash your car, then you of course must not expect that these very same people will then connect with you. Or if you try to link up with them, then they of course look away.... That has nothing to do with the fact that that person has a different faith, but with the fact that he does not want to adjust his behaviour to the culture in that village. Those who publicly show no respect for this.... That is then irritating of course. And then you must not complain that you are not accepted, because that will then never happen.

An important aspect in the acceptance of migrants in the local community is thus that their behaviour must not contradict the dominant view and valuation on how daily life in the local community is organised. Conation is more important than affection in this matter. Crucial in the relations between the mostly non-religious migrants and some of the local communities is thus not so much religiosity but public behaviour. In communities like Ouddorp, Sunday rest has become part of a local identity discourse, which is also accepted by non-religious groups. Non-religious natives are better positioned socially and cognitively to deal with this than non-religious migrants who are not familiar with the social norms governing the local way of life. A non-religious woman who was born and bred on Goeree-Overflakkee explains:

I always experienced it personally as mutual respect. I come from the island, so I have easier access than when you come completely from the outside and you don't speak the language. I rented for years a small house when I was still single from people from a very orthodox church. Sometimes they said: "oh, you go on holidays on a Sunday?" He did not approve of that he told me. But I always went to him for a coffee on Sundays; just mutual respect. They said that they deplored I did not go to church. But they were kind to me and we were just close with each other and still are. I never experienced malice or something like that.

Box 5.3 The relation between church communities

Figure 5.2 New orthodox church in Ouddorp.
Source: photo taken by Maarten Hogenstijn.

These issues related to religiosity and respect were frequently raised during the interviews in reaction to our showing of this picture of a new church in Ouddorp (Figure 5.2). These and other new churches were recently built on the island as a consequence of the recent breakaways of orthodox-protestants from the main protestant church in the Netherlands. This division influenced the daily life in many local communities. Although in Ouddorp the vast majority of the parish split away from the main church, the church building remained in the ownership of the few who stayed in the main protestant church. This forced the thousands of breakaway orthodox-protestants to look for new accommodation. After worshiping in temporary accommodation like sports halls, they quickly built a large new church at the edge of the village, as depicted in Figure 5.2. According to one of our interviewees, the volume of these new churches built for the breakaway orthodox-protestant groups had outstripped the volume of buildings constructed for businesses on the island in the last five years. Those interviewees who were directly and personally affected by this spilt were quite hesitant to discuss it. They stress the painful fresh fissures now running through their local community and which some-times even divide families.

However, it hardly affects the political position of orthodox-protestants in local politics. The party (SGP) which represents their interests had always incorporated people coming from many different orthodox-protestants churches, which were mostly formed after tumultuous breakaways. Religion is not a topic in this party. They respect or ignore their religious differences. They focus instead on the articulation of their shared interests of religious groups in the municipality with regard to issues like Sunday rest, subsidies for religious schooling, and housing and employment for their large families with many children (Hogenstijn & van Middelkoop 2008).

A non-religious member of a local village council who moved almost 40 years ago to the island and who is actively attached to the local community reacted to this picture as follows:

> Yes religion, yes, for the majority of our island it also signifies a traditional conservative attitude in many situations. I find it very awkward, this picture symbolises religion on our island, but that has also many different aspects. Well, if I look at this church, then quite honestly, it is not something which makes me happy, because it is a very extreme corner within our society. Yes and they also just cause irritations, not only with me. We just have to deal with various sections of the community and you all try to appreciate each other. I think everybody really reckons with Sunday rest which is important for the religious parties on this island. But I don't experience that this respect is mutual. Yes, let me emphasise again that that only a part is so religious. We have here six churches in Ouddorp, and some of them deal with it in different ways. There are also here boys from families who go to church on Sundays, who also work here on Sundays, no problem. They also go on Sunday, when the weather is nice, to the beach and to a beach pavilion, that is not a problem. But some people won't allow anything on a Sunday. I understand that we should not go to the village centre making a racket, but for instance we had recently a sales weekend here for our new project. On Saturday somebody stopped by to tell us what fools we were to be open on Sunday to sell these bungalows. And I think, we do this at the edge of the village, somewhere in a meadow, where we won't disturb anybody at all. We did not do it in the village as we respect you. Then don't come lecture us but respect us too!

5.5.1 The zoning of Sunday rest in the old municipality of Goedereede

Spatial zoning is an important element in dealing with different perspectives on Sunday rest. In the tourist village of Ouddorp, all shops and most of the pubs and restaurants in the village centre are closed on Sundays. There are also no outdoor activities there. Further away from the village centre, the situation is different. The large holiday parks just outside the village have their own shops, which are open on Sundays. Further away, on the Brouwersdam connecting Goeree-Overflakkee to the next island, there are even large outdoor festivals on Sundays. This dam has always belonged to the old municipality of Goedereede, but the local population does not regard this dam, which was built in the 1960s, as part of their village. In the local mental map, it is part of a gradient from the village where they live. The further away from the village centre, the more is allowed. On the fringe of the village, where the holiday parks provide many of them with an income, facilities are open on Sundays. On the Brouwersdam, much more is allowed. Even further away, the touristic village of Renesse, located at the other side of a sea arm, is seen as the dystopia of a village taken over by hordes of drunken and promiscuous youths from Rotterdam and its urbanised region. People from Ouddorp fear that further development of local tourism will transform them into a similar Sodom and Gomorrah. Or in the words of one of our interviewees: "Renesse is just a madhouse."

The different views and interests in relation to Sunday rest in the old municipality of Goedereede were accommodated in the municipal politics. The development of tourism, which generates a lot of income for many religious families, was never opposed even by the political party of the orthodox-protestants. Its development was, however, controlled with the feared dystopia of Renesse literally looming over the horizon. This control not only involved the very visible zoning of touristic activities described above, but also the size and type of accommodation was controlled. Large-scale developments were avoided as much as possible and located far away from the village centre. Some of our interviewees informed us of the subtle ways in which the issue of Sunday rest was addressed for these newer developments. In the planning phase, Sunday rest was always an issue raised by local opponents. Developers then typically agreed to close their facilities on Sundays. After the construction of these new resorts, the local population got used to them and realised that it was not as bad as Renesse and provided jobs for the local community. Then, typically, the company concerned contacted the municipal administration and argued that their closure on Sundays threatened their profitability and employment opportunities. The municipality then normally would quietly agree to their opening up on Sundays. These practices were part of the consensus-based local politics in the old municipality of Goedereede. In the words of a leading politician of the orthodox Christian political party:

> The identity here is more conservative. The social-democratic (PvdA) councillors would back us in like 90 per cent of the issues. While now, it is difficult to put a figure on it, it is more like 60 per cent. That differs

and that has to do with how a local community is composed. Look, the father of a leading social-democratic politician was an alderman of our party, you see? Thus those lefties are different in such a closed community on this part of the island compared to the rest of the island. Sure, also people from non-Christian political parties in the former municipality of Goedereede, people mainly from Ouddorp, they understand possibly better how a local community in Ouddorp functions, than people from non-Christian political parties in Middelharnis.

5.5.2 Dealing with local sensitivities after amalgamation

The amalgamation of Goeree-Overflakkee changed the established local political landscapes. The background of the resistance from the population and municipality of Goedereede against amalgamation has already been discussed in §5.1. The feared loosening of the regulation of Sunday rest was part of this resistance. Although, after amalgamation, the local resistance identity discourse has more or less disappeared and become overshadowed by a thinner regional identity discourse focussing on the economic development of the island, the issue of Sunday rest has become a bone of contention in the local politics of the amalgamated municipality.

A local leader of the orthodox-protestant party discusses how the political parties in the municipal council functioned in the old municipality of Goedereede and compares this to the amalgamated municipality of Goeree-Overflakkee:

> They tolerated each other, they accepted and understood each other's positions, the parties on the left and right. The Christian parties agreed: "no shops open in the centre", and the parties on the left: "yes, but then open on the village edge". The polder model worked optimally here in Goedereede. (...) Now in the amalgamated municipality politics is conducted on the cutting edge. Left and right is now the dividing line. Thus the polder model has vanished. That does not mean that is wrong because you can also be too nice to each other.

Another local resistance leader against the amalgamation of Goedereede comments:

> Local politicians were really afraid that the identity of the former municipality Goedereede would be lost through municipal amalgamation. I think that they might be right, because when you see how the municipal council now operates, they would not have done that in the former municipality Goedereede based on their identity.

A different local political leader of the orthodox-protestant party fears that the issues they regarded as crucial for their identity and which they had previously protected in their former municipality are now unsecured on the slippery slope of the politics of the new municipality, which they can no longer control.

Many of our interviewees on the western part of the island, especially those who opposed amalgamation, combine this fear of losing control over issues they relate to their local identity with financial issues. In their opinion, the taxes their tourist trade generates for the amalgamated municipality are used to solve the financial problems created by the former municipalities on the eastern part of the island. On the other hand, some of our interviewees on the eastern part of the island feel neglected. The policies of the new municipality focus, in their opinion, too much on the western touristic part of the island, while they get the things nobody wants, like huge wind turbines.

On the eastern part of the island, the Christian parties were too weak to strictly enforce Sunday rest. The shops were closed, but there was ample room for other activities on Sunday. For instance, a local baseball team had its fields right next to an orthodox church. Sometimes they played tournaments on Sundays. An inhabitant explains how this potential conflict was resolved informally in this local community.

> When they had, for instance, Sundays a tournament then they asked from the church if it was possible to start after one o'clock, after church service. Well, that was just done like that. The music was put on only after one o'clock. That was how that was resolved.

This changed after the amalgamation. The municipal administration now tries to apply a single policy on Sunday rest. It is not just the orthodox-protestants on the western part of the island who fear that this will undermine Sunday rest; many liberals on the eastern part also fear that this will limit their freedom in how they conduct their daily life. In the summer of 2014, this resulted in a conflict after the municipality forced a travelling circus to cancel its performance on Sunday. A national newspaper quoted a local councillor living in the eastern part of the island:

> Perhaps some people find it exaggerated, but if the SGP wants to pursue at any price their views on Sunday rest for everybody on the island Goeree-Overflakkee, then I say: "the caliphate Goeree-Overflakkee is coming, if it is not already there yet".

(AD 2014)

For this comparison between the policies of the municipality and a caliphate, he got support from his side of the island, but angry reactions mainly from Christians from the other side of the island. The fear of a religious caliphate was invoked after the travelling circus, which had always given shows in Oude-Tonge on the eastern part of the island every summer for a whole week, ran into problems for the first time when they applied for the necessary permissions from the new municipality. The municipal administration pressurised the circus into not performing on Sunday. After agreeing to this, the circus got its licence. According to the municipality, the circus voluntarily agreed not to perform on Sunday, but its manager claimed they were forced to agree. This is resented by large sections of the local community in Oude-Tonge, since the travelling circus had always performed on Sundays in the past, before the municipal amalgamation.

Our interviewees come up with many other topics on which the different opinions about the role of religion in society have generated conflicts in the municipal council of the amalgamated municipality. These range from municipal facilities like swimming pools, the upkeep of cemeteries, praying at the beginning of council meetings, support for Christian schools and associations, the opening hours of shops and the organisation of events like cycling races and travelling circuses.

5.5.3 Conclusion: all lose their dominant positions after amalgamation

Many (see also Chapter 1) have concluded that after municipal amalgamation local politics becomes more confrontational. This is linked to the increased size of the municipality, which results in a lack of knowledge and empathy for the interests and identity of the many different local communities in an amalgamated municipality. This is also related to the growing distance between the amalgamated municipality and the population. The previous discussion of how the old and new municipality deals with issues on Sundays rest showed that size involves more than population numbers. The spatial differentiation on the island was also important. First of all, amalgamating the more orthodox communities in the west and the more liberal communities in the east resulted in a loss of influence in municipal politics for both. They both lost their local political dominance and thus felt aggrieved after amalgamation. Besides this quantitative loss by the dilution of local power in a wider arena, local politics also experienced a qualitative change. In each former municipality on the island, local political parties had over time developed their own way of dealing with different visions on local identity, like, for instance, issues related to Sunday rest. This established secondary identity of dealing with different primary local identities and different views on this disappeared through amalgamation. The construction of a new, broad secondary identity is difficult. Goeree-Overflakkee was relatively successful in constructing a new thinner regional identity discourse focussing on the economic development and marketing of the island. Creating a similar consensus on more cultural issues, like the regulation of Sunday rest, proved however to be more difficult.

References

AD (2014). Op weg naar een kalifaat op Goeree-Overflakkee. *Algemeen Dagblad*, 9 August, page 29.

Goedereede (2010). *Zienswijze in het kader van het Herindelingsontwerp Goeree-Overflakkee van Gedeputeerde Staten van Zuid-Holland van 15 maart 2010.* Retrieved from: https://herindelingnee.wordpress.com/zienswijze-gg/.

Hogenstijn, M. & D.P. van Middelkoop (2008). *"Zo werkt dat hier niet": gevestigden en buitenstaanders in nieuwe sociale en ruimtelijke kaders.* Delft: Eburon.

IPGO (2011). *Aangenaam Goeree-Overflakkee: voor een ambitieus en duurzaam economisch beleid in de 21e eeuw.* Middelharnis: Innovatie Platform Goeree-Overflakkee.

Jeekel, P., D.O. de Leth, M. Nijhout & B. Snoeker (2013). *Een onderzoek naar verzet tegen gemeentelijke herindelingen.* Utrecht: honours leeronderzoek SGPL.

Rabobank (2006). *Kiezen of delen: op zoek naar een duurzame balans op Goeree-Overflakkee.* Middelharnis: Rabobank Goeree-Overflakkee.

Rabobank (2014). *Samenwerken aan een vitaal en duurzaam Goeree-Overflakkee – nu doorpakken voor een sterk en vitaal eiland.* Middelharnis: Rabobank Goeree-Overflakkee.

6 Katwijk

"A city which has remained a village"

Katwijk took the lead in the common struggle of more than 20 years against urbanisation, which is the main threat hovering over the region. That threatened its identity, its small-scale character, its identifiability. The flower and bulb growing industry, the commerce, the whole card house would collapse. There is here so much individuality and pride and so on, for which they started to fight more than 20 years ago.

(Regional administrator)

Being different from their urban neighbours is an important element in the local identity discourses in both Katwijk and Goeree-Overflakkee. The general role of anti-urbanism in how our interviewees perceive their local identity was discussed in §4.8. This chapter analyses how local identities were used in the political struggle against urbanisation in Katwijk. It starts by analysing how the local politics in Katwijk and its neighbouring rural municipalities reacted against the planned building of a large new town in their region in the 1990s. This resulted in the development of a regional identity discourse through which local politicians in Katwijk and in the municipalities of its rural region joined forces with local agricultural entrepreneurs, conservationists and the provincial administration. Together they successfully opposed the large-scale urbanisation planned by the central government. The emergence of this regional identity discourse and also its current fading away are discussed in §6.1. Attention then shifts in §6.2 to how local identities were used to oppose another, much more localised urbanisation threat in the 2000s. The regulation of this new urbanisation threat was an important reason why, in 2006, three municipalities decided to amalgamate. Their main aim was to be stronger together in order to protect their distinct local identities against external threats. After amalgamation, the different local identities became a key element in the local politics (§6.3). In §6.4, the role of local neighbourhood councils is analysed. This is followed by an analysis of the role of local associations in the persistent use and dominant role of local identities in the municipal politics of Katwijk.

6.1 Katwijk and the ring of Dutch cities

Katwijk is located on the fringes of the Dutch urban core, the ring of cities which is known as the Randstad (see also Figure 3.2). In §4.8, we discussed the fact that, although its high population density puts Katwijk, according to official statistics, into the category of an urban municipality, our interviewees regard it as "villages welded together" or as "a city which has remained a village". Even though it is surrounded by the cities of the Randstad, the dominant view in Katwijk is that they are not part of the Randstad.

The Randstad started in the 1950s to become a key planning concept to regulate post-war urban growth in the Netherlands. The urban policies of the central government wanted to prevent the creation of an amorphous huge city. One set of regional economic policies focussed on reducing the migration from rural areas into the Dutch cities in the Randstad. Another set of spatial policies aimed to control urban sprawl by keeping nearby rural spaces open. These green spaces should prevent cities in the western part of the Netherlands coalescing into a single mega-city. These open spaces were to be used for recreation and would protect areas with highly specialised agriculture. Katwijk was considered a part of the Leiden agglomeration, which was separated from the nearby Haarlem agglomeration by an open area with dunes for recreation and the agricultural fields in the Bollenstreek specialised in bulb and flower cultivation (Beenakker 2008; Duineveld & Van Assche 2011).

6.1.1 The rural identity of the Bollenstreek threatened by large-scale urbanisation

The Bollenstreek, stretching northwards from Katwijk, was until the nineteenth century a sparsely populated infertile dune landscape (see also Figure 3.2). During the nineteenth century, the inland dunes were dug off to supply sand for the expansion of the nearby cities. The remaining wastelands were re-cultivated at the end of the nineteenth century. The chalk-rich and well-drained sandy soils were highly suitable for the cultivation of flower bulbs (Beenakker 2008; Duineveld & Van Assche 2011). The dominance of bulb growing gave this region its name and shaped the identity of the Bollenstreek. This rural regional identity discourse was not based on a long tradition of agriculture, but based on the modernity of controlling nature. The soil was made by man through re-cultivation. The production of flower bulbs and flowers for decoration in urban homes was an agricultural innovation. Both were important elements in a more or less modernist regional identity discourse. This corresponded to the general pride of the Dutch of their conquest over nature, which was an important element in the Dutch national identity discourse. That element was downloaded from the national to the regional level. (In §2.7 the importance of downloading in the construction of layered spatial identities was discussed on a more abstract level.) The bulbs, and especially the tulip, became the core of a regional identity discourse. The tulip also became an important symbol in the representation of Dutch national identity. This regional identity element was thus uploaded to the national level. After the Second World War, the growing demand for bulbs, requiring larger and more efficient fields, caused bulb production to spread to other areas in the Netherlands. Furthermore, the expansion of nearby cities slowly eroded the area used for bulb production in the Bollenstreek. Today only 10 per cent of Dutch bulbs are produced in this area, but the trading companies located in the Bollenstreek control 80 per cent of the world trade in flower bulbs (Duineveld & Van Assche 2011; Terlouw & van Gorp 2014).

This restructuring of the agribusiness sector and the decline of bulb production coincided with an increase in the urbanisation pressures from the nearby booming Amsterdam Schiphol airport region. Young, emerging bulb growers especially felt increasingly threatened by urbanisation. With the help of the local branch of the main association of agricultural entrepreneurs, they started to cooperate from the end of the 1980s with other entrepreneurs, such as the bulb traders, to better collectively protect their interests. In the 1990s, the threat of incremental urbanisation from existing towns within the region was surpassed by the plans of the central government to build a huge new town of more than 100,000 inhabitants. This would more than double the regional population. The declining importance of the bulb fields made the Bollenstreek look empty from a national perspective and therefore suitable for relieving the growing urbanisation pressures from the nearby booming Amsterdam Schiphol airport region and the city of Leiden. Urbanisation now threatened not only individual bulb growers, but the entire bulb industry and the entire regional structure of the Bollenstreek. Although the old bulb fields are no longer crucial for the production of bulbs, the agribusiness complex needs room for the expansion of offices, distribution centres and other facilities. The planned large new town would hinder this (Duineveld & Van Assche 2011; Kloosterman 2001; De Vries & Evers 2008; VROM 2007; Terlouw & Weststrate 2013).

The organisations of the bulb and flower agribusiness sector were part of a regional coalition of different stakeholders, uniting municipalities, environmentalists and the provincial administration; this alliance successfully opposed the planned new town. The regional coalition against the national plan for a new town did not use a purely defensive strategy. This would not fit with the growth orientation of the bulb growers' network. It would also have reduced the bargaining position of the Bollenstreek towards the growth-oriented national government. Their strategy was not just to block national urbanisation plans, but to transform them in alignment with the interests of the local stakeholders to create new opportunities for regional development. The regional identity discourse they developed not only opposed large-scale urbanisation, but also linked up with different aspects of other more established spatial identities. By doing so they prevented the national government from downscaling the conflict by branding the local protest a NIMBY driven by special interests (§4.5.4). Instead they successfully upscaled the conflict (§4.5.3).

In their opposition to this new town, the environmentalists in particular communicated a rural identity of the Bollenstreek based on downloading elements of new national policy discourses on landscape, heritage and ecology. The bulb fields close to the dunes were represented as a unique Dutch landscape; old barns were rebranded as bulb heritage to be protected for demolition. Even some species of field birds were rebranded as "bulb birds". These and other local groups were supported by the municipalities in the Bollenstreek. By participating in the resistance to urbanisation, these municipalities broadened the scope of their cooperation in the Duin- en Bollenstreek, a new functional region which was initially created to collectively provide public services for the participating municipalities. The province of Zuid-Holland also opposed the proposed large-scale urbanisation in the Bollenstreek since it identified the region as having a unique landscape, whose ecosystem needed improvement through the renaturation of the old bulb fields. It also helped those who opposed large-scale urbanisation to link up with the stakeholders of coastal tourism because renaturation of bulb fields would increase the touristic potential of the region. This regional opposition was further supported by the rural resistance identity against urbanisation used by the inhabitants and the municipalities.

6.1.2 From resistance against urbanisation towards a Greenport

This cooperation between these different regional stakeholders against this large new town was based on a shared vision on the future of this region. This was first laid down in 1996 in the *pact van Teylingen*. While this document focussed more on restrictions than on developments, the pact was extended in 2004 into the more development-oriented *offensief van*

Teylingen. These policy documents focus on spatial zoning, planning procedures and many specific projects agreed upon by all the partners involved: the province of Zuid-Holland, the municipalities in the region, their associations, the district water board, environmental protection associations and agribusiness associations. They all agree on the central position of bulb growing in the region.

> Bulb growing gives the region its specific identity, the bulbs are an iconic part of the landscape and attract millions of tourists yearly. The bulb growing agribusiness complex is the back bone of this region for its economy, communities, culture, history and landscape.
>
> (Stuurgroep 2002, 21)

In a later document, the association of local entrepreneurs named "Keep it flowering" further developed this vision on the region (Stichting Hou Het Bloeiend 2006).

Box 6.1 "Keep it flowering"

The regionally rooted but branching-out network of the bulb and flower agribusiness complex dominates this spatial conceptualisation of the Bollenstreek. The agribusiness activities dominate the landscape and are rooted in the soil type and accumulated local knowledge. They form a regionally integrated network which is firmly nested within the regional territory, but which expands with ever more important international linkages, facilitated by the nearby Amsterdam Schiphol airport and the port of Rotterdam. This makes the region very important for the competitiveness of the Dutch national economy. The flowering bulb fields in spring also attract many international tourists. The regional economic development and landscape are dominated not only by the production of bulbs and flowers, but more and more by related activities, like research, education, marketing, distribution, inspection services, biotech laboratories, international trade and transport services. This creates many connected and dense networks which dominate and fill out the regional territory.

In order to strengthen its dominant international position in the production and trade in bulbs – and to a lesser extent flowers – the use of the regional territory has to be restructured to make it more sustainable for this agribusiness complex. Space has to be made available for more high-quality bulb production, trade park extensions and a more attractive landscape for both inhabitants and tourists. The linkages between these and other more localised activities in the region integrate its territory. The expanding agribusiness is seen as an indication of the strength of this regionally rooted but expanding network. The further development of the regional economy needs a restructuring of land-use and improvements in the local and regional infra-structure. A multitude of very specific projects were accordingly formulated in cooperation with local and regional administrations and other regional stakeholders. For their realisation, they support the formation of a semi-autonomous public–private partnership (PPP) based regional development company (Stichting Hou Het Bloeiend 2006; Stuurgroep 2002).

The association of local entrepreneurs intensified the cooperation with the regional municipalities and in 2007 they created the Greenport Duin- en Bollenstreek. This further strengthened the spatial conceptualisation of the region as a network.

> The Greenport is a place without borders. We don't discuss borders, we are bound by the boundaries of the six municipalities, but others are not excluded from Greenport activities. We are discussing the Greenport, thus we consider all the people who shape the Greenport and the companies involved. Then you don't consider municipalities, at least not with regard to borders.
>
> (Regional administrator)

A Greenport is an agribusiness complex which the national government considers to be important for the global competitiveness of the Dutch economy. They have a similar function as the Mainports Amsterdam Schiphol airport and the Rotterdam harbour (VROM 2006). The national government stimulates the development of Greenports by easing planning regulations, improving infrastructure and stimulating cooperation between companies to innovate. This cooperation further intensified in 2010 with the founding of a regional development company by the municipalities in the Duin- en Bollenstreek, which implements the common spatial policy based on this Greenport concept and related visions discussed in Box 6.1 (Ecorys 2006; SDB 2001, 2003a, 2003b; WLTO/KAVB 2003).

A regional administrator comments on the role of the regional Green-port development company in the period after the successful resistance against the building of a large new town:

> One is here of the opinion that we have won from the central state; it can remain open. But if we do not open it up genuinely, and let the economy function there, then the urbanisation threat will return. Therefore the Greenport development company has been given room to quickly and energetically open things up and restructure and so forth.

Another regional administrator comments on the role and function of the Greenport Duin- en Bollenstreek for the regional economy as a whole.

> It really is about the strengthening of the Greenport to keep it here, to attract subsidies, to be known and market the region through the shared implementation of economic policies. These are very broad, including region marketing. The positioning of the Greenport in the market is the figurehead. But it is also about strengthening its base. Keeping bulb growers in the region, retaining know-how. This is the cradle of bulb growing, this is where it originated. Now it is cultivated everywhere, in the Flevo polder, in the North, much more than here. Only the knowledge comes from here. We have to think hard about that, how to retain this? How to embed this in the region and use it?"

Later he comments on some recent research into the economic strengths of the region: "Surprisingly these economic reports show that the residential attractiveness is very important here. Why do you think all those people want to live here? Of course due to the landscape. So the landscape determines the residential attractiveness."

6.1.3 Beyond the Greenport

The business associations were very successful in influencing local and regional policies and in promoting their economic interests. In cooperation with other stakeholders, in the late 1990s they averted the threat of large-scale urbanisation of the region through the planned construction of a large new town. The Greenport development company also successfully restructured large parts of the obsolete agricultural landscape with its old sheds and greenhouses in the early 2010s. The main infrastructural bottlenecks were also fixed. Most of the projects formulated at the beginning of the cooperation between these different stakeholders in the region are now realised or are in the process of being realised by the Greenport development company.

Their successes have undermined the basis of this Greenport cooperation within the regional borders of the Duin- en Bollenstreek. Both the municipalities and entrepreneurs cooperate in ever more networks, but with less and less territorial fixity. Municipalities cooperate less with other municipalities on a multitude of topics in the same region. They increasingly cooperate on different topics with different municipalities outside the region. The administrators of Katwijk quite openly pursue a regional geopolitical strategy based on cooperating on different topics with many different partners. They are afraid to cooperate too frequently with the same partners. They fear that this will result in the formation of a strong regional organisation. This will threaten the independence of Katwijk and the identity of its local communities. They also fear that this could over time even result in further amalgamations. The administration and local politicians of Katwijk are not against regional cooperation, but only want to cooperate when it is directly beneficial to Katwijk. Cooperation is for them not a goal in itself. "Katwijk is like a lone wolf in the greater whole. They of course participate out of necessity, but there is no identification or natural feeling of solidarity" (local inhabitant).

Box 6.2 Functional overlapping competing jurisdictions (FOCJs)

Bruno Frey (2005) sees FOCJs as an alternative to the hierarchical administrative system of a centralised state.

While traditional administrative levels are hierarchically organised and controlled from above, FOCJs emerge from below in response to the geography of specific problems. FOCJs can more effectively and efficiently deliver public goods. This is not an alternative, but an additional administrative structure. Municipalities are the most common building blocks of these FOCJs. These are *functional* regions where the area covered is defined by the specific task they have to fulfil. Their size fits the scope of the interests of the stakeholders. As different functions cover different areas, they *overlap* not only because in one area several regions operate with different functions, but also because different FOCJs with the same function, and catering for this function, need not be contiguous and may (partially) overlap. They are *competing* because participants can choose to some degree their involvement with specific FOCJs. There are (democratic) mechanisms through which they can voice their interests or they can exit from it by not using their services. They are *jurisdictions* because they implement governmental powers, policies, projects and regulations.

Frey conceptualises them as an ideal typical alternative to which specific examples only comply to a certain degree. Thus while ideally FOCJs would finance themselves through levying their own taxes, in praxis new regions depend very much on centralised funding and contributions from the participant administrations (Frey 2005).

Many of our interviewees link the decline in cooperation between the municipalities in this region to the recent threat of further municipal amalgamations in the Netherlands (see also §1.1). A multitude of only partially overlapping and constantly changing new regions has emerged resembling more and more FOCJs (see Box 6.2). For instance, Katwijk and Noordwijk cooperate in garbage collection and taxation; Katwijk, Noordwijk and Teylingen together maintain their infrastructure; Noordwijk, Noordwijker- hout, Lisse, Hillegom and Teylingen cooperate in social services; Katwijk cooperates with Wassenaar and Oestgeest for the care of disabled people. The focus of cooperation also shifts away from the Bollenstreek to Leiden and its surrounding region. "Within many municipalities you get com- plaints of the patchwork of collaborations. But that does not matter in my opinion. You must detach that from those borders, think in networks. That works best over here" (regional administrator).

The local entrepreneurs also focus their cooperation less and less on the Greenport Duin- en Bollenstreek. Their attention has widened to the Green- port Holland, an association of all six Greenports in the Netherlands. The entrepreneurs also broaden their cooperation outside the agribusiness sector through, for instance, cooperating more with the Leiden Bio Science Park and the Holland Space Cluster in Leiden. But their focus has also narrowed towards more specific forms of cooperation, like the Flowers Science Centre in Lisse. The agribusiness networks thus re-spatialise away from the coher- ent single and territorially fixed Greenport Duin- en Bollenstreek. Or in the words of a regional administrator: "I suspect that the Greenport will be par- tially abandoned, only the connections will be retained." Another regional administrator gives her view on the growing fragmentation of this region:

> The cooperation in the region suffers from this enduring discord. This is not only affecting the municipalities – the region is notorious for this – but is also noticeable in the business world. This hinders getting everybody on one side. This burdens themselves the most. We notice frictions between the older and the younger generations. Perhaps not so much frictions, but a different perception of how to deal with things. They cannot link up together, in my opinion. The old genera- tion is committed to the cooperation in the Bollenstreek region and want to preserve this. The new generation wants to profit from other opportunities. This is difficult to bridge. It also depends on the com- plicated situation in relation to the self-centred municipal strife. There is also a lot of discord within local business associations. Discord characterises this region. This has to do with the fact that the municip- alities are of the same size, without a leader taking the initiative. This is the same in the business world.

In 2014, an economic board of entrepreneurs and academics which did not have direct links with the business associations was formed, partly to overcome the divisions that have emerged recently in the regional business community.

6.1.4 The thinning and fading of the regional identity discourse

The opposition to a new town in the region generated a regional identity discourse combining elements of a territorially thick rural resistance identity and a thinner regional identity based on the agribusiness complex. The regional identity discourse was strengthened by the selective up- and downloading of elements of related spatial identity discourses, which were linked to the different stakeholders in the regional coalition against the new town. This use of a regional identity discourse which combined elements of an established thick rural identity with elements of a competitive thinner economic identity cemented a strong coalition which successfully blocked the building of a large new town in this region. After this urbanisation threat was averted, the regional cooperation and the regional identity discourse transformed. In 2004, the cooperation network of the municipalities merged with the neighbouring HollandRijnland cooperation network, which has a more diffuse identity discourse than the Bollenstreek. The municipalities of the Bollenstreek also cooperate in the Greenport Duin- en Bollenstreek. This now communicates a narrower and thinner identity discourse, which is predominantly based on their expanding agribusiness network and its links to the Dutch economy. The specific characteristics of the territory of the Bollenstreek have become, over time, less important in this dominant regional identity discourse. The threatened large-scale urbanisation cemented a strong regional coalition focussing on the protection of the regional territory against unwanted developments. This was helped by a regional identity discourse linking thicker territorial regional elements with thinner outward-oriented economic elements. After averting the threat of the new town, the focus shifted ever more from territorial defence to network development. This initially focussed on the Greenport, but this focus is now fading. It has more or less completed its policy agenda, and is becoming overshadowed by many other partially overlapping networks of entrepreneurs, municipalities and other government agencies.

6.2 The amalgamation of Katwijk and the fear of neighbouring Leiden

Urbanisation did not stop in the Bollenstreek after the successful opposition against the planned large new town. Gradual urbanisation and migration towards Katwijk and its neighbouring municipalities have always continued from the nearby urban areas. This gradual urbanisation is based on the development of new housing estates ranging in size from a few to a few hundred houses. The scarcity on the housing market which drives this urbanisation is not only caused by migration from nearby cities and the attraction of the nearby booming Amsterdam Schiphol airport area. The building of some new houses is also considered necessary to provide housing for particularly young members of the local communities in Katwijk. By slowing down and reducing the size of the new housing estates, municipalities try to control the inflow of urban migrants. The gradual development of small housing estates better matches the limited but steady local demand for new housing. In contrast, the rapid construction of huge housing estates would flood the local housing market, with too many houses to be filled by local people. This would then result in a huge influx of unwanted urban migrants.

The urbanisation threat increased again when the national government announced that the military airfield in the rural municipality of Valkenburg was to close in 2006 and that its area was to be redeveloped into a huge housing estate with many thousands of houses. This was opposed by the neighbouring local communities, especially since the neighbouring city of Leiden (122,000 inhabitants) wanted to annex Valkenburg to develop the old military airfield as its new suburb. The inhabitants and administration of the rural municipality of Valkenburg (3,900 inhabitants) feared losing their local identity. Or as one local politician formulates it:

> The threat came in those days from Leiden. Valkenburg feared losing control to Leiden. We must guard that Leiden builds its houses on the airfield. Then they will swallow us as well. That did not happen and I think they were haunted by the fear to become part of Leiden. That threat was averted.

Valkenburg successfully allied itself with the old municipality of Katwijk (43,000 inhabitants) and Rijnsburg (15,000 inhabitants), which also feared the urban expansion of Leiden. Through the amalgamation with Valkenburg, they wanted to control and reduce this housing development on their doorstep.

The voluntary amalgamation in 2006 of the old municipalities of Katwijk, Rijnsburg and Valkenburg into the new municipality of Katwijk, with now 65,000 inhabitants, makes it big enough to resist further amalgamations and control external influences. The vision document on the future of the new municipality of Katwijk reads like a declaration of independence. "The amalgamation of Katwijk, Rijnsburg and Valkenburg in the unitary municipality Katwijk is a step towards independence" (Katwijk 2007, 13).

The new municipality stresses the individuality of the three old municipalities and their desire for local self-determination. Amalgamation made

Katwijk a stronger player on the regional playing field, dominated by mounting external pressures from increased urbanisation and regional cooperation. The protection of common anti-urban interests was an important driving force in the amalgamation process. This amalgamation of quite similar local communities was a pragmatic second-best solution, which was preferable over annexation by Leiden, amalgamation with dissimilar municipalities like the tourist resort Noordwijk, the sub-urban municipality of Oestgeest, or losing control to regional organisations promoting the development of the wider metropolitan region. The municipalities Katwijk, Valkenburg and Rijnsburg therefore decided in 2002 to merge in order to protect their different local identities against these common external threats. They also developed a common vision on the amalgamated municipality and its spatial development strategy covering the period until 2020. This was based on the results of many sessions with different types of local stakeholders organised by consultancy firms. This vision was laid down in a document titled "A sea of possibilities" (Een zee aan mogelijkheden) (Katwijk 2007). In this vision, the amalgamated municipality of Katwijk was represented as a mosaic of different local communities and areas. This diversity had to be protected, but also connected by infrastructure improvements and by the joint development of waterfronts along the Rhine, which runs through the whole amalgamated municipality of Katwijk and connects all its different villages. This shared river is also the reason why, when in 1999 the local branches of the Rabobank in Katwijk, Rijnsburg and Valkenburg merged, they chose the name "Rhine villages" (Rijndorpen) for the amalgamated local branch. This new name was, however, not adopted by the new municipality, because Katwijk, the largest merging partner, opposed it. Another indication of the difficulties in creating a new identity for the amalgamated municipality is that almost all our interviewees did not recognise the officially adopted picture of Katwijk as a mosaic city which was presented in this official common vision document as the "ideological guideline" (Katwijk 2007, 11) for the amalgamated municipality. One of the few local politicians who recognised this mosaic city concept comments: "Yes, the spatial development strategy! This can still give me nightmares. There were so many discussions, comments, reports, all those consultancy firms involved. And then in the end, what was the result of all that planning ..."

In Katwijk, the amalgamation of municipalities was strongly linked to the protection of local identities. The new municipality focusses on the preservation of the different thick local identities and actively opposes the development of a distinct identity for the new municipality as a whole. Whereas on Goeree-Overflakkee a thinner complementary overarching identity is seen to support local identities, in Katwijk an overarching identity is regarded as a threat to the different local identities. In Katwijk, local identities are used to stress the differences between local communities within the amalgamated municipality.

> If you talk with a Rijnsburger he will tell you that the village of Katwijk is the last place on earth he will want to live. That is the old rivalry between the communities of Katwijk and Rijnsburg. Valkenburg is still a village on its own, really a close-knit village community.

6.3 After amalgamation: "own village first"

The local politics in Katwijk after amalgamation is characterised by an inhabitant as based on an "own village first" mentality. A divisive municipal politics dominates in Katwijk, which focusses on the equal distribution of municipal investments and services over the different local communities. The establishment of neighbourhood councils after amalgamation is a clear expression of the dominance of local identities and the concern for the equal distribution in the municipal politics. Their role is discussed in the next section. The many active local associations are, however, much more important in this divisive use of local identities. In the last sections of this chapter, we study the role of the local Orange associations for the use of local identities in the amalgamated municipality of Katwijk.

This divisive type of local politics is further entrenched by the functioning of the local political parties. Local communities do not have their own political party, but every political party is very aware of the importance of selecting candidates from each local community.

> That is taken into account by the putting together of the list of candidates for the municipal elections. Certainly. We did not have until late an inhabitant of Valkenburg on the list. We searched high and low, looked in the smallest corners until we found one. That was an important consideration. The election result, you can verify it, in the voting districts, I think 90 to 95 per cent of the votes went to members of that local community.
>
> (Local politician)

The members of the different local communities are very attentive to the fair distribution of investments and services of the new municipality over the different local communities. "What we very often hear is that when the municipality has a project, it must be divided among the different places. People keep arguing that they also want what others have, and that they don't want to be disadvantaged" (local administrator). This type of discontent with the spatial distribution of municipal investments and services was only mentioned once or twice on Goeree-Overflakkee. It was, however, a prominent theme in most of our interviews in Katwijk. Members from all different local communities in Katwijk expressed that their community was disadvantaged by the new municipality. This is very similar to the views of the population in amalgamated rural municipalities in Australia. Every local community claims that they get less than the other local communities in the amalgamated municipality. This coincides with an increased level of distrust between neighbouring local communities after amalgamation (Alexander 2013).

6.4 Neighbourhood councils

Dealing with the different local identities after municipal amalgamation was an important topic during the political discussions on amalgamation. It was therefore decided that five neighbourhood councils would be established. Not only did the old municipalities of Rijnsburg and Valkenburg both get a neighbourhood council, but the two villages in the old municipality of Katwijk – Katwijk aan Zee and Katwijk aan de Rijn – also got their own neighbourhood council, as well as the post-war neighbourhood Katwijk Noord, since this area has a similar number of inhabitants as the two villages.

> Neighbourhood councils were set up by the municipality. Neighbourhood council members are also appointed by the municipality. The purpose was in reaction to the amalgamation, to preserve the identity of the places. That's why neighbourhood councils were established. The intention was that neighbourhood councillors, who could self-apply, have strong roots in the community, based on associations, church or whatever.
>
> (Community worker)

These neighbourhood councils act as an intermediary between the local population and the municipal administration. They mostly deal with issues related to objects in public spaces affecting the daily life of inhabitants. Neighbourhood councils typically receive complaints from individuals about the construction or removal of things, like parking places, bollards and trees, and pass these on to their contacts in the municipal administration. The neighbourhood councils we interviewed were predominantly occupied by these kinds of mundane issues. Although these neighbourhood councils were known by our interviewees, they were never discussed as a key element in the relation between local identities and politics. Some even regard it as "a kind of extra unnecessary administrative level which only slows things down". The local population has an ambivalent attitude towards neighbourhood councils. One neighbourhood councillor told us that in a survey 96 per cent of the population attached great importance to the neighbourhood council, but that never more than a few inhabitants went to their meetings. Another neighbourhood councillor remembers that they once had a meeting about the reconstruction of an important road. He expected that this locally important issue would attract more people than usual. He expected a few dozen:

> but 600 came. We had to relocate the meeting to the church. Thus if there are important issues they will come (...). After the meeting they gave us a standing ovation, but nobody approached any members of our neighbourhood council and asked if they could participate and help us. You see? If others pull the cart it's fine and you're a great guy. But if you do something wrong, then they complain.

The neighbourhood council members of Valkenburg illustrate their troubled relation with the amalgamated municipality of Katwijk by recounting a dispute over the planting and uprooting of a few dozen trees. These trees were planted as part of a new neighbourhood park, which was conceptualised by an external consultant. These trees only survived after planting because a local inhabitant took the trouble to water them. Then after some time the municipality decided to cut these trees down for the construction of a football field. The neighbourhood council wrote to the municipality and asked them to replace these trees. This was, however, interpreted by the municipality not as a friendly suggestion, but as an official complaint and thus became part of legal procedures. In the end, the municipality refused to replant all the trees since it did not comply with the masterplan developed by the architect they had employed. The members of the neighbourhood council:

> talked so long that their tongues were almost blistered. We discussed and discussed it for two whole afternoons and now in the end they planted 25 trees. For that we had to move heaven and earth and thank the good Lord. But there was room for more than 50 trees. You see, you must struggle to get things done. You have to be on top of it, time and time again, you have to remind them constantly otherwise nothing ever happens. You are just trivialised. That's how it is, that's how we feel it.
>
> (Neighbourhood councillor)

The legitimacy of the neighbourhood councils is not based on elections, but on the involvement of its members with the local community. There is an implicit expectation that members should belong to a well-established local family or they should be actively involved in the local community. Members are expected to have a strong identification with their community, they must be traditionally rooted, but they can also be actively attached (see §4.3). Many members of neighbourhood councils frown upon new inhabitants who want to join their councils in order to become part of the local community. Length of residency is thus a key criterion for becoming a legitimate member of a neighbourhood council. There appears to be a kind of ideal typical member of a neighbourhood council, whose family has always lived there and who has always been actively involved in local community life. In Katwijk our interviewees generally regard the neighbourhood council of Valkenburg as an example of a strong council, partially because their members are so strongly rooted locally. However, when we interviewed members of the neighbourhood council of Valkenburg, even those who had lived their whole life in Valkenburg downplayed their nativeness. A true Valkenburger appears to be a perfection aspired to, but which can never be completely achieved. The members of the neighbourhood council in Valkenburg who are considered by the rest of the local communities in Katwijk as true native Valkenburgers downplay their nativeness themselves. We interviewed them as a group and they reacted to the question of what constitutes a true Valkenburger in terms of their length of residency. The ones who lived in Valkenburg their whole life started the discussion. They downplayed their nativeness by focussing on Leiden as their place of birth, since Valkenburg lacks a hospital, or that they could have with hindsight moved to another part of the Netherlands, or that even though they lived here their whole life, they were in their younger years not actively involved in the local community. A neighbourhood councillor claims he is totally not a Valkenburger even though he has lived there for 22 years and has always been actively involved in the local community. He was reassured by a fellow neighbourhood councillor that being actively involved in the local community is decisive. "But what is a genuine Valkenburger. I find that hard to define. But you are a Valkenburger if you are really involved in the community."

A group of neighbourhood workers in Katwijk described the backgrounds of the neighbourhood councils in the old municipalities of Valkenburg and Rijnsburg:

> They started out as action groups, as protest movements. In the beginning they were very activistic. They were more action committees than neighbourhood councils advising the municipal administration.

They wanted to show Katwijk that one cannot annex Valkenburg or Rijnsburg just like that. That is reflected by the swiftness in which they attracted council members, and still they attract new members quite easily in contrast to other neighbourhood councils (...). When you look at individual neighbourhood council members, in my opinion, many of them become members out of a general sense of discord or some specific dispute. We try to transform that hostility into positive energy.

Another community worker commented: "Every neighbourhood council wages its own battle with the municipality."

Neighbourhood councils are very active in promoting the particularistic spatial interests of their community to the municipality. Neighbourhood or village councils are often instituted in the Netherlands and abroad by amalgamated municipalities to reduce the gap between population and administration and to protect the different local identities of the amalgamated places. Amalgamation processes are frequently opposed by the population based on the feared loss of their distinct local identities by the amalgamation into a monolithic municipality. How this can result in the formation of resistance identities has been discussed in §2.8 and §5.1. Many amalgamated municipalities hope to appease the population of old municipalities, which lost their municipal council, by establishing neighbourhood or village councils. As the above case of Katwijk suggests, the establishment of neighbourhood councils can on the contrary not diminish but institutionalise conflicts. Especially since, contrary to the old municipal councils they replace, neighbourhood or village councils cannot directly address or redress local problems, but have to convince the amalgamated municipality to take action. This often results, as discussed above, in recurrent conflicts between the neighbourhood council and the amalgamated municipality. To strengthen their case towards the municipality and to mobilise local support, these neighbourhood councils tend to focus on the threats to their local identities by the actions of the amalgamated municipality, which more or less institutionalises local resistance identities.

Before amalgamation, individual citizens could relatively easily articulate their interests and concerns directly to the municipal administrations. This was especially the case in the smaller municipalities, where inhabitants, local politicians and municipal officers were very familiar with each other and their interests. A local entrepreneur observes:

The distance between the municipal administration and the citizen increases. Especially in Valkenburg. There you could, so to speak, without an appointment, walk into the office of the alderman and mayor. It was very informal there. In Rijnsburg it was largely the same. That distance is now much bigger. Everything is now decided in the faraway buildings of the amalgamated municipality and we have to endure it. Look, politics represents all places. Local interests are adequately represented in the municipal council. But fewer and fewer municipal officials are linked to the local communities. These have in a way no longer feelings for local issues and identities.

Local interests of citizens are now articulated in a more indirect and confrontational manner. For instance, through their local representatives in municipal council or their neighbourhood council, aggrieved citizens can mobilise support from within their local community to force the municipality to address their interests. This strengthens the importance of local identities, which are reformulated and used in different ways. Instead of informal individual contacts and cooperation, local politics is more based on the confrontation between local communities which strengthens and thickens their different local identities. In the old municipalities, local identity helped to integrate local communities. In the new municipality, local identities are used in the confrontation between local communities. The increased distance between the population and their administration after amalgamation is observed in many different countries (Hansen 2013; Mecking 2012; Ruggiero *et al.* 2012; Baldersheim & Rose 2010). The development of a more confrontational local political system focussing on the fair distribution of investments and services in the amalgamated municipality appears to go hand in hand with the growing discontent of citizens. This was also the case in German municipalities after amalgamation, as discussed in Box 6.3.

The establishment of neighbourhood councils was in Katwijk directly linked to the public debate about amalgamation. They are still the most visible manifestation of the use of the different local identities in municipal politics. They are part of an institutionalised municipal political culture or secondary identity which focusses on the differences between local communities and the jealous guarding of a fair distribution of public goods and municipal investments over the different local communities. However, they are more a visible manifestation than the driving force behind the use of local identities in the relations between local communities. When we discussed the role of identities in local politics with our interviewees, it was not the neighbourhood councils but the wide array of local associations that were considered crucial. There are many active associations in Katwijk. Compared to the rest of the Netherlands, these local communities have a very active social life. This is mentioned by all our interviewees in Katwijk. They also stress that, although almost everybody participates in local associations, almost all

---

Box 6.3 The legacy of the German amalgamations of the 1960s and 1970s

In the 1960s and 1970s, West Germany had a major administrative restructuring. The number of municipalities was reduced by 65 per cent, from 24,278 to 8,514. In the federal state of Nordrhein-Westfalen, the reduction was 83 per cent. Like after other amalgamations, the increased distance between population and administration can be linked to the lack of identification with the new amalgamated municipality. This was, by many involved in this process, regarded afterwards as one of the main drawbacks of these amalgamations (Mecking 2012).

For instance, the merger of the two small towns of Villingen and Schwenningen in the federal state of Baden-Würtemberg did not create a new municipal identity, but in fact strengthened the different local identities of both towns. This amalgamated municipality was the result of a marriage of convenience. It became the central town in this area where the central government invested heavily in new services like education, infrastructure and recreation facilities in the amalgamated municipality. Their different local identities did not diminish, but were strengthened by the amalgamated municipality through subsidising the different local associations (Reuber 1999).

In Wattenscheid in the Ruhr area, in the 1960s there was very strong opposition against the amalgamation with the larger town of Bochum. The entire municipal council voted against this merger. Even in 1996, decades after the merger, 87.7 per cent of the population still wanted to undo this merger (Mecking 2012). The enduring strength of their local identity is reflected in their appreciation for their own car number plates. Before 2012, all car number plates in Germany started with an abbreviation of the municipality of the car owner. The loss of municipal autonomy was thus reflected by the number plates on your car. When that regulation changed in 2012, dozens of inhabitants of Wattenscheid stood for hours in the freezing cold to be the first to change their number plates starting with the hated BO(chum) with their beloved WAT(tenscheid) (Arendt 2012).

---

are only members of the associations of their own community. Each local community has, for instance, its own highly competitive amateur football club. Choral societies are also widespread, but very locally organised and very competitive. This is even the case for Orange associations, which support the Dutch royal family. The most active Orange associations in the Netherlands are found in the municipality of Katwijk, but each local community has its own Orange association organising huge local popular festivals on different dates and in different places. These Orange associations are considered by our interviewees to be the pivots around which the social life of the different local communities is organised. These Orange associations play a crucial role in the institutionalisation of the different local identities in the amalgamated municipality of Katwijk. The next sections analyse in more detail the development of these local Orange associations and their role in the institutionalisation of local identities.

6.5 Local Orange associations: one nation, four villages

You wonder why four Orange associations in one village? That was always so in history. Katwijk is the place to be on Queen's Day. Katwijk aan de Rijn celebrated the birthday of Wilhelmina on 31 August. That harmonised with harvest time. Katwijk aan de Rijn is more agriculture, horticulture. The harvest is then almost over, some money was earned, enabling them to booze for a few days. Valkenburg has the oldest horse fair of the Netherlands. That is the focus of the festival week. And Rijnsburg has its flower parade for many years tied in with a traditional popular festival. Thus every place has its own festival, its own identity. It's impracticable to merge, then they must organise four festivals; all festivals of about a week. That is unmanageable of course. Everybody is committed to its own identity.

(Chairperson of local Orange association)

This quote from a chairperson of one of the four Orange associations in the municipality of Katwijk introduces the paradox which is examined in the last sections of this chapter. The four very active Orange associations are united in their traditional royalist version of Dutch nationalism. They are, however, divided in their expressions and celebrations of their local identities. Despite their shared focus on expressing their traditional national identity, these Orange associations focus on their own festivals celebrating their local and national identities, which are separated from each other not only in space, but also in time.

The situation in Katwijk appears to contradict the dominant view of the succession of local identities by national identities. With the development of nation-states, nationalism is supposed to replace localism as the main focus of identification, uniting previously rivalling communities. The doctrine of nationalism stresses the homogeneity of the nation and the exclusivity of the relation between the individual and the nation (Smith 1982; Yack 2012, 107; Billig 1995). Recent discussions on the renegotiating of nationalism and the linked rise of local and regional identities present this predominantly as a zero-sum game based on the mutual decline of the national and the rise of the local (Flint & Taylor 2007). Many argue, however, that the identification with the nation was more a normative ideal than an empirical reality.

The idea of "identity", and a "national identity" in particular, did not gestate and incubate in human experience "naturally", did not emerge out of that experience as a self-evident "fact of life". That idea was forced into the Lebenswelt of modern men and women – and arrived as a fiction. It congealed into a "fact", a "given", precisely because it had been a fiction.

(Bauman 2004, 20)

The relation between local and national identities is more complex than a simple succession from the local to national. Local and national identity discourses are linked to each other in a constantly changing political process through which both are changed. National identities are built on a mosaic of traditional regional and local identities. Nation-state building involved the institutionalisation of distinct local and especially regional stereotypes based on tradition and fixed in distinct regional territories. The national identity was conceptualised as crowning these traditional regional identities and uniting them by providing them with a collective path to a better future. Regional identities based on naturally bounded historical regions were especially seen as important for the formation of active national citizens (Paasi 2012, 6). In this nation-focussed hierarchical identity discourse, local and regional identities are thus reduced to the remnants of the past, while the nation is linked to the future. Local and regional identities are thus transformed through the growing importance of national identities. This link between the traditional past and local and regional identities imposed by nationalistic discourses can also explain why local identities are, by many, still regarded as remnants of the past. Box 6.4 discusses how in Germany the idea of Heimat bonds the local, regional and national identity discourses in Germany.

Box 6.4 The nation and local identity: the German Heimat

The Heimat idea yearned for the past not because it was antimodern but because it originated from modernity.

(Confino 1997, 156)

Heimat is a key concept in studying the formation of a national identity discourse in Germany after its unification in 1871. Germany was a diverse amalgamated territory, with federated states which had a very diverse political history merged into a single political entity. Its borders also excluded large sections of the old German empire and the German-speaking world. German unification also coincided with the transformation of German society through industrialisation and urbanisation.

In this situation, the local identity discourse based on Heimat emerged, which focussed on the historical roots of the now changing everyday local life. The concept of Heimat was at first used by local elites in their opposition to modernisation which threatened the traditional rural way of life. The local community was used as the core symbol and was linked to security, harmony, stability, timelessness, daily life, mother and home (Confino 1997). Initially, it stressed the need for the protection of local heritage sites and landscapes threatened by industrialisation and urbanisation. Later on, Heimat focussed more on the local and regional differences of German modernisation. After the end of the nineteenth century, the German way of dealing with modernity became an important component of an increasingly dominant conservative interpretation of German nationalism. The German nation was conceptualised as a mosaic of different Heimats, where the general German culture was rooted in historically grown, specific local and regional identities. The initial opposition between the local Heimat and Germany's national development was transformed by using these Heimats in the construction of a German national identity. The concept of Heimat was used to articulate a distinct national path to modernity in opposition to other especially Western European nations (Confino 1997; Applegate 1990; Cremer & Klein 1990; Spohn 2002; Conze 2005).

The Heimat was thus not linked to a specific scale level, but linked to the different but similar local and regional historical roots of the national path to a better future.

> Germans manufactured Heimat as a set of shared ideas about the immemorial heritage of the German people in local and national history, nature, and folklore as, first of all, part of a European and North American response to modernity. (...) As one of Germany's responses to modernity, the Heimat idea was a memory invented just when German society was rapidly changing, as a bridge between past and present that looked uniquely dissimilar. Heimat looked to the past for reassurances of uniqueness on the local and the national level in times of political, economic, and cultural homogenization: it emphasized the uniqueness of a locality with respect to national standardisation, and the uniqueness of Germany with respect to European and North American standardization.
>
> (Confino 1997, 98)

The discourses on Heimat and local and regional identities have changed since the end of the nineteenth century. During the Nazi regime, Heimat was reformulated in terms of race, blood and soil. This fascist legacy discredited the Heimat idea in the first decades after the Second World War. In the 1960s, Heimat became part of more left-leaning identity discourses and was used as a counter-symbol to West German society with its renewed nationalism, American consumerism and environmental pollution.

> For more than a century, then, the Heimat idea served as a metaphor for the changing relationship between localness and nationhood in German society.
>
> (Confino 1998, 193)

6.6 The history of the Orange associations in Katwijk

In the late nineteenth century, the Dutch government dominated by liberals tried to develop a nationalistic spirit because they felt threatened by the growing unrest among the lower classes. However, in 1868, 1872, 1873, 1876, 1879 and 1884, the official celebrations of important events in Dutch national history failed to attract much popular support (Huijsen 2012, 72, 85–86). In 1885, not the government but local newspapers campaigned to publicly celebrate the fifth birthday of Princess (after 1890, Queen) Wilhelmina, who was the heir to the throne of the unpopular King William III, who died in 1890. Her birth was framed as a "blossoming flower on a languishing Orange family tree" (Micklinghoff 2010, 19). Queen's Day had, from the start, more the character of a popular festival than an official celebration. It was organised locally and focussed very much on the involvement of the common people. Since 31 August was at the end of summer, Queen's Day was also used to replace traditional local harvest festivals with this celebration of national unity. Located at the end of the school holidays, it also focussed from the start on the involvement of the youth (Micklinghoff 2010, 39–41).

In Katwijk aan Zee, the birth of Princess Juliana in 1909 led to the establishment of an Orange association to organise the festivities which for many years had been organised by local schools. The character of the Queen's Day celebrations focussed initially very much on the involvement of local children.

On 4 September 1913, the Orange association organised an independence festival, celebrating the centenary of the liberation from French Napoleonic rule. Like in many other places in the Netherlands, they re-enacted local

Table 6.1 Royal holidays in the Netherlands

1885–1890	31 August	Princess's Day (Wilhelmina)
1890–1948	31 August	Queen's Day (Wilhelmina)
1948–1980	30 April	Queen's Day (Juliana)
1981–2013	30 April	Queen's Day (Beatrix)
2014–	27 April	King's Day (Willem-Alexander)

Source: Micklinghoff (2010, 19–21).

events related to the end of French occupation and the founding of the Kingdom of the Netherlands. Over the years, it became a local tradition to celebrate the independence from French rule through the re-enactment of the coming ashore of the later King William I on the beach. This actually took place at another fishing village on 30 November 1813 and is therefore a good example of an invented tradition. This propagated a very traditional Orangist anti-liberal version of Dutch nationalism, as it linked the ideas of national independence with the opposition to the ideals of the French Revolution. This celebration of national liberation became part of the Queen's Day celebrations in Katwijk aan Zee. After the occupation by the Germans in the Second World War, the focus of this celebration of national liberation shifted from the liberation from the French (and enlightenment) to the liberation from the Germans (and fascism).

The remembrance culture linked to the Second World War is very important for the creation of a Dutch national identity, which had been relatively weak until then (van Ginkel 2004; Lijphart 1968). The celebration of national independence also shifted after the Second World War to 5 May, bringing it together with the Queen's Day celebrations for the new Queen Juliana on 30 April, as part of a week of celebrations of Dutch national identity. In recent decades, street decoration using orange and national colours has remained in place to support the Dutch national football team when they play in European or World Championships. Similar to other celebrations of independence, like those in Finland, it shows that:

> once accomplished, (political) independence is not merely a fixed state of affairs but a dynamic, contextual, and contestable process that positions the state as part of the ceaselessly shifting geopolitical landscape and can take on innumerable local and national, and even contradictory forms in material landscapes, practices and narratives.
>
> (Paasi 2016, 30)

These types of changes are taking place not only at the general national level, but also at the local level in Katwijk. The incorporation of local festivals, re-enactments, remembrance, and involvement of the common people and the youth are all elements which we also find in other local festivals celebrating nationalism (Confino 1997).

6.7 The four different local Orange associations

In every place there are active large Orange associations. They always compete for prominence. Competing Orange associations. Rijnsburg has a very large one. So large that it can compete with Katwijk aan

Zee, which has its own Orange association and also Katwijk aan de Rijn has its own Orange association. But these are huge Orange associations. That they have in common, royalism.

(Local politician)

The Orange association in Katwijk aan Zee is one of the oldest (1910) and largest (4,000 members from a local population of 14,000) in the Netherlands. They are very proud of that and boast their special ties with the Orange family. In 1916, the royal family chose the tranquillity of Katwijk for a beach holiday. The later Queen Juliana lived in the traditional village of Katwijk aan Zee in the period 1927–1930 when she studied in urban Leiden. There have also been several royal visits to Katwijk aan Zee in the past few decades. In 2000, the Queen visited the Queen's Day festivities in Katwijk aan Zee. Their prominence and special linkage to the house of Orange is exemplified in the large-scale festivities they organise. Around King's Day, they now organise a multitude of activities with a focus on the youth and the local population. These include traditional games, sports matches, music performances and a large funfair. The organisers claim that 80–90 per cent of people put out their flag on King's Day. About 50,000 to 60,000 people attend these festivities (Micklinghoff 2010).

Many members of especially the orthodox-protestant sections of the local community do not want these festivities to expand further. They fear that a further expansion could transform it into a hedonistic large festival like in other more urban coastal places. This is also a major reason why the Orange association of Katwijk aan Zee refused the request of the municipality to participate in the revitalisation of the mainly touristic fishing days festival. They wanted instead to concentrate on their own festival. "Everybody has its own festival supporting separate identities" (board member of local Orange association).

The other Orange associations were established later: in Katwijk aan de Rijn in 1913 (currently a membership of about 2,300 members of 8,000 inhabitants), Rijnsburg in 1945 (about 2,200 members from a population of 15,000) and Valkenburg in 1931 (about 750 members from a population of 4,000). They always had difficulties competing with the larger Orange association in Katwijk aan Zee and their established large-scale festivities around Queen's Day. In the late 1990s, issues related to local identities became important topics in local political debates about municipal amalgamation and urbanisation. This coincided with an increase in the competition between the different Orange associations. In 1997, the Orange association of Katwijk aan Zee was temporarily relegated into second place in membership numbers because the Orange association of neighbouring Rijnsburg had managed to increase its membership to 3,035, outnumbering the 1,876 members in Katwijk aan de Rijn. This motivated the Orange association in Katwijk aan Zee to increase its activities (Micklinghoff 2010). This further stimulated the other Orange associations to focus even more on their local Orange festivals. Due to the established dominant position of the Queen's Day and later King's Day festivities in Katwijk aan Zee, the activities organised by other Orange associations on this day attract few people. This motivated the other Orange associations to develop their own local Orange festivals in other periods. The differences between the local Orange associations and their local festivals thus increased at the same time as the external urbanisation threat to local identities became an important political issue.

Orange associations not only organise the birthday celebrations of the monarch, but are also involved in many different local festivities. Some of these are national occasions like the feast of St Nicholas, or New Year; others are traditional local festivals. In Katwijk aan de Rijn, the Orange association continued organising festivities in the period around the birthday of Queen Wilhelmina at the end of August after her abdication in 1948. Initially the reason was that, contrary to the fishing village Katwijk aan Zee, Katwijk aan de Rijn was an agricultural village. The horticulturalists in Katwijk aan de Rijn did not have the time for huge celebrations in the spring. Queen's Day had become part of their local autumn harvest festival. This fall festival expanded over time. Initially it included a small-scale music festival in a schoolyard, but this has expanded enormously, especially since the 1990s. In that period, the Orange association in Katwijk aan de Rijn could not compete anymore with the festivities at Queen's Day in Katwijk aan Zee, which has also included since 1989 a large funfair. In that period, protecting local identity also became a political issue linked to the urbanisation threat of the planned large new town which was discussed earlier in this chapter. Since the 1990s, the festival in Katwijk aan de Rijn has continued to expand. It was drawn-out from a Friday, to a weekend, to a full week. "You don't want to know what large events they come up with. They really go to extremes. They are supported by many active volunteers" (local administrator). Presently it attracts close to 100,000 visitors over the week, has 150 separate events and frequently stages Guinness Book of Record bests.

The festival site in Katwijk aan de Rijn is, however, allocated for housing development. It is still a wasteland due to the collapse of the Dutch housing market after the financial crisis of 2008. Since their site is threatened by urbanisation, they are looking for a new and preferably larger festival site. This is very difficult because almost all the space within and near the village has been built over in the last few decades. That is one of the reasons why many in Katwijk aan de Rijn want to put a noisy road into a tunnel. This road carries mostly regional traffic and also divides their village. While the local and provincial administrations have invested heavily over the last few years to reduce the problems created by roads in other villages, they now feel that they are also entitled to their own tunnel. This would also make room for a large, centrally located festival space where the local Orange association would have more room to organise their big autumn festival.

The Orange association of neighbouring Rijnsburg envies them:

Katwijk aan de Rijn is in my experience a very close-knit community, based on its Orange association. If they organise something everybody participates. We are very jealous of that. We ask before the festival week for people to put up orange garlands, we even distributed them. Then some come, but with 200 garlands you cannot decorate a whole village. But if you visit Katwijk aan de Rijn during their festival week they decorated the entire village. They have a very large group of volunteers, we are very envious of that. But it shows what a resilient community they are, who do a lot of things together.

(Board member of Orange association)

The Orange association in Rijnsburg focusses their activities on festivities related to the yearly flower parade at the beginning of August. The flower parade takes place during the festival week in Rijnsburg, which is organised by the Orange association, but the flower parade itself is not officially organised by the Orange committee. This celebrates the successful local economy of Rijnsburg, which is based on the flower trade. "Flowers is their pride. Their market hall, the flower parade, they very much want to identify with those, and with the related business mentality, that's what they regard as the genuine character of Rijnsburg: merchants" (community worker). The flower parade is an expression of their local identity, which is also very outward oriented. They use the English name "flower parade" to attract foreign tourists. The flower parade attracts about 250,000 visitors yearly. The organisers of the flower parade are also involved in the organisation of flower parades in other countries, like Azerbaijan.

The flower parade is the grand finale of the local festival week in Rijnsburg. Although the flower parade attracts a lot of outside visitors, the whole festival week is important in bringing together the local community.

It attracts present and past inhabitants of Rijnsburg. There are also people who especially get here for this occasion. Especially Thursday, the biggest feast day, then almost everybody takes a day off. And on

Friday the flower market hall closes early. Many arrive then early in the village. The children's flower parade departs from the market hall, with all children on their bikes decorated with flowers. It first crosses the market hall area, so it attracts the attention of the people working there. You hear that people arrive on Thursday to visit mom and dad. They are then on the loose for the entire day. It is really like long time no see. A kind of reunion. Also when you retire from work you are invited for coffee and cake. We will give you a present. The other day I had this mail from someone: "I am a Rijnsburger, but I live since thirty years in the Province Drenthe. Can I come?" Of course you are welcome, fantastic. There are really people who plan their vacation around it.

(Board member of Orange association)

Since 1931, Valkenburg has had its own local Orange association. They organise some small-scale celebrations on King's Day and Remembrance Day, but their main event is the yearly festival week organised around the horse fair in September. It boasts being the oldest in the Netherlands, dating back to 840. On the opening page of their website, they write: "We continue to fight to protect our own identity, thus one Valkenburgse Orange association for all Valkenburgers!"

The horse fair is regarded as a fair for the local population. Some people from neighbouring villages visit it, but only fleetingly; it is really a social event for native Valkenburgers. Unlike the other villages, they do not use the national flag and orange decoration, but they use extensively their local yellow-red flags. Their festivities exemplify traditional Dutch village life.

But Valkenburg, nice locality, we have a traditional bonfire of Christmas trees, St Nicholas still arrives by boat and with black Pete's, we walk during the horse fair with a beer in public, without being fined, that is just special. That sounds strange, but it is the amiability of a village. And if my neighbour gets drunk we bring him home. That is real village life. It sounds perhaps corny, but ...

(Local businessman)

The amalgamation had some negative effects on their local festivals. A few regret that the beer stops flowing after 1:00 am. Others felt threatened by the attempts made by the amalgamated municipality to transform some local events into get-togethers for the whole municipality.

The municipality of Katwijk tries to develop King's Day with municipal subsidies as a collective festivity. But no Valkenburger comes there. A bit exaggerated, but just a few dozen at most. Valkenburg organises something locally. Even Remembrance Day, the focus is on Valkenburg as always. With our 10 May unfortunate history, it is our Remembrance Day. It is not of Katwijk, nor of Rijnsburg. It is our Remembrance Day. When that was to change suddenly, when the amalgamated municipality Katwijk wanted to change that, people revolted almost. That is not possible. It is our Remembrance Day, it has to take place here! Also, in the end the municipality announced that the Remembrance Day ceremony would take place in Valkenburg, because of the historical events in Valkenburg and which are linked to this commemoration. The battle over Valkenburg airfield. That lasted a year [author note: not the military battle, but the dispute over the remembrance ceremony]. Then everybody went its own way. That does not make you happy. No, I cannot think of something we share.

The same traditional royal family oriented brand of Dutch nationalism unites all four Orange associations, but they celebrate their local Dutch national identity on separate occasions. This diversity has increased over the past decades. Only after the municipal amalgamation in 2006 have they occasionally cooperated. The amalgamated municipality supports an association of all four Orange associations in the municipality. They have organised together festivities for two special occasions: in 2010, the 65-year remembrance of liberation, and in 2013 during the enthronement of King Willem-Alexander. Besides these exceptions, all these very active Orange associations organise their distinct local festival week and many other local events. All year round they organise smaller local festivities, on King's Day, Liberation Day, the entry of St Nicholas (with black Pete), New Year's Eve, etc. However, only Katwijk Noord has no Orange association. There is no local festival, only some smaller-scale street barbecues and Sugar Feast celebrations.

A different culture really. In terms of identity, comparatively especially with the large neighbouring cities like Leiden you can characterise them easily. You look for similarities between three or four smaller communities. You are inclined to do that. These have few differences, but these are identifiable. There are cliques in every place that want to preserve it. Katwijk aan Zee say don't touch our Queen's Day, that is ours, we are proud of it. Valkenburg horse fair. Katwijk aan de Rijn has its autumn festival. All say we have the greatest festival. And honestly as a Katwijk Zeeër, I have never visited the horse fair in Valkenburg, nor the flower parade in Rijnsburg, I have seen it pass, but.... Every village has its own stuff. But that stuff ... those places, they are ours. For an outsider these four Orange associations could easily merge. No, impossible. They can find each other for the enthronement or royal wedding. Then there are huge events sponsored by the municipality. That goes well on such occasions. But otherwise, all apart.

(Local businessman)

6.8 The unity and diversity in Katwijk's identities

The threat of urbanisation to local identities has dominated local politics in Katwijk over the past decades. Urbanisation threatens them internally and externally. Internally, the expanding fringe neighbourhoods with many urban and foreign migrants challenge their traditional way of life, almost from within. Externally, they experienced different urbanisation threats. In the 1990s, they worked closely together with other communities in their rural region to oppose large-scale urbanisation and created a strong regional cooperation. After this threat was averted, this regional cooperation itself came to be seen increasingly as an external threat to local autonomy. Later, there was also a more localised urbanisation threat from the nearby city of Leiden. To protect their local autonomy, they merged into a municipality big enough to resist urbanisation and in which no local community could dominate the others. The balance between the four local communities is the basis for a local distributive local politics based on the mobilisation of local dissatisfaction over distribution of municipal services and investments. Local neighbourhood councils are a clear expression of this municipal political system dominated by discontent and conflicts over distribution.

The four local Orange associations are key players in the institutionalisation of these four distinct local communities. They also bolster the local identity discourses based on local differences and rivalry, which are however rooted in a shared traditionalist version of Dutch national identity, which opposes urbanisation and migration. All four Orange associations are important in the expression and celebration of both local and national identities. They express a very similar vision of Dutch national identity and how this is related to traditional Dutch village life in opposition to urbanisation and migration. They get their strength not only from this shared national traditionalistic village idyll, but also from their local rivalry, which is embedded in a divisive municipal political constellation.

The dominant local identity discourse based on local differences and autonomy created a kind of secondary identity discourse which helped them internally to cooperate and externally to communicate a distinct anti-urban identity. Their Orange associations provide them with a hold they can safely cling to in their hostile environment with the pressures from urbanisation, migration and rival communities threatening their traditional way of life.

References

Alexander, D. (2013). Crossing boundaries: action networks, amalgamation and inter-community trust in a small rural shire. *Local Government Studies*, 39, 463–487.

Applegate, C. (1990). *A nation of provincials: the German idea of Heimat*. Berkeley: University of California Press.

Arendt, N. (2012). *WAT-Kennzeichen feiert erfolgreiche Rückkehr*. Ruhr Nachrichten, 14 November.

Baldersheim, H. & L.E. Rose (Eds) (2010). *Territorial choice: the politics of boundaries and borders*. New York: Palgrave.

Bauman, Z. (2004). *Identity: conversations with Benedetto Vecchi*. Cambridge: Polity.

Beenakker, J. (2008). *1000 jaar Duin- en Bollenstreek*. Zwolle: Waanders.

Billig, M. (1995). *Banal nationalism*. London: Sage.

Confino, A. (1997). *The nation as a local metaphor*. Chapel Hill: University of North Carolina Press.

Confino, A. (1998). Edgar Reitz's Heimat and German nationhood: film, memory, and understandings of the past. *German History*, 16, 185–208.

Conze, V. (2005). *Das Europa der Deutschen: Ideen von Europa in Deutschland zwischen Reichstradition und Westorientierung (1920–1970)*. München: Oldenbourg Verlag.

Cremer, W. & A. Klein (Eds) (1990). *Heimat: Analyses, Themen, Perspektiven*. Bielefeld: Westfalen Verlag.

De Vries, J. & D. Evers (2008). *Bestuur en ruimte: de Randstad in internationaal perspectief*. Den Haag: Ruimtelijk Planbureau.

Duineveld, M. & K. Van Assche (2011). The power of tulips: constructing nature and heritage in a contested landscape. *Journal of Environmental Policy & Planning*, 13, 79–98.

Ecorys (2006). *Greenport Duin- en Bollenstreek: analyse + uitvoeringsagenda 2006–2020*. Rotterdam: Stichting Hou het Bloeiend.

Flint, C. & P. Taylor (2007). *Political geography: world-economy, nation-state and locality*. Harlow: Pearson.

Frey, B.S. (2005). Functional, overlapping, competing jurisdisctions: redrawing the geographical borders of administration. *European Journal of Law Reform*, 5, 543–555.

Hansen, S.W. (2013). Polity size and local political trust: a quasi-experiment using municipal mergers in Denmark. *Scandinavian Political Studies*, 36, 43–66.

Huijsen, C. (2012). *Nederland en het verhaal van Oranje*. Amsterdam: Balans

Katwijk (2007). *BSV Katwijk. Een zee aan mogelijkheden. Brede Structuurvisie 2007–2020*. Katwijk: gemeente Katwijk.

Kloosterman, R.C. (2001). Clustering of economic activities in polycentric urban regions: the case of the Randstad. *Urban Studies*, 38, 717–732.

Lijphart, A. (1968). *The politics of accommodation: pluralism and democracy in the Netherlands*. Berkeley: University of California Press.

Mecking, S. (2012). *Bürgerwille und Gebietsreform: Demokratieentwicklung und Neuordnung von Staat und Gesellschaft in Nordrhein-Westfalen 1965–2000*. München: Oldenbourg Verlag.

Micklinghoff, F.H. (2010). *Koninklijk Katwijk: 100 jaar oranjevereniging Katwijk aan Zee*. Den Haag: Euromedia.

Paasi, A. (2012). Regional planning and the mobilization of "regional identity": from bounded spaces to relational complexity. *Regional Studies*, 47, 1206–1219.

Paasi, A. (2016). Dancing on the graves: independence, hot/banal nationalism and the mobilization of memory. *Political Geography*, 54, 21–31.

Reuber, P. (1999). *Raumbezogene Politische Konflikte. Geographische Konfliktforschung am Beispiel von Gemeindegebietsreformen*. Stuttgart: Franz Steiner Verlag.

Ruggiero, P., P. Monfardini & R. Mussari (2012). Territorial boundaries as limits: a Foucauldian analysis of the agglomeration of municipalities. *International Journal of Public Administration*, 35, 492–506.

SDB (2001). *Vijf jaar Pact van Teylingen: een toekomstperspectief voor de Duin- en Bollenstreek*. Lisse: Samenwerkingsorgaan Duin- en Bollenstreek.

SDB (2003a). *Integrale gebiedsanalyse handhavingssamenwerking regio Duin- en Bollenstreek*. Lisse: Samenwerkingsorgaan Duin- en Bollenstreek.

SDB (2003b). *Offensief van Teylingen*. Lisse: Samenwerkingsorgaan Duin- en Bollenstreek.

Smith, A. (1982). Ethnic identity and world order. *Millennium: Journal of International Studies*, 12, 149–161.

Spohn, W. (2002). Continuities and changes of Europe in German national identity. In: M. Malmborg & B. Stråth (Eds) *The meaning of Europe: variety and contention within and among nations*. Oxford: Berg.

Stichting Hou Het Bloeiend (2006). *Greenport Duin- en Bollenstreek: een vitale economie in een vitaal landschap*. Lisse: Stichting Hou Het Bloeiend.

Stuurgroep (2002). *Vijf jaar pact van Teylingen: een toekomstperspectief voor de Duin- en Bollenstreek*. Lisse: Stuurgroep Uitvoering Pact van Teylingen.

Terlouw, K. & J. Weststrate (2013). Regions as vehicles for local interests: the spatial strategies of medieval and modern urban elites in the Netherlands. *Journal of Historical Geography*, 40, 24–35.

Terlouw, K. & B. van Gorp (2014). Layering spatial identities: the identity discourses of new regions. *Environment and Planning A*, 46, 852–866.

Van Ginkel, R. (2004). Re-creating "Dutchness": cultural colonisation in post-war Holland. *Nations and Nationalism*, 10, 421–438.

VROM (2006). *Nota ruimte: ruimte voor ontwikkeling*. Den Haag: VROM.

VROM (2007). *Strategische toekomstagenda Randstad 2040: naar een duurzame en concurrerende Europese topregio*. Den Haag: VROM.

WLTO/KAVB (2003). *Regiovisie Duin- en Bollenstreek*. Haarlem: WLTO/KAVB.

Yack, B. (2012). *Nationalism and the moral psychology of community*. Chicago: University of Chicago Press.

7 Conclusion

The divergent use of local identities

This book has shown that local identities can play an important role in local politics. It has analysed how local and regional identity discourses have changed in two municipalities and how this is related to shifts in local politics. There are plenty of other cases in the Netherlands and elsewhere where identity discourses play an important role in political relations. These are not limited to municipalities and their regions as studied in this book, but are also present at and between other scales. These range from the relation between neighbourhood communities (Elias & Scotson 1965), to nationalism (Paasi 1996), the relations between nation-states and the European Union (Delanty & Rumford 2005) and the contacts between Western and other civilisations (Said 1978). This book did not analyse the general use of identity in politics, nor did it determine the overall importance of identities in these political relations. The aims of this book were more modest: exploring the different ways in which local identities are used in politics.

This book showed the nuanced and varied use of local identities in local politics. The index at the end of this book refers for instance to over 20 different aspects of identity. Local identities are used in politics in many different ways to negotiate the old and the new.

As discussed in Chapter 1, identities are seen by most politicians, administrators and academics as obstacles in political processes. Rearranging territories and the powers between spatial scales tend to bring identities to the fore. The detailed analyses in this book showed how local identity discourses in both Katwijk and Goeree-Overflakkee were adapted to changing circumstances. Based on their experiences on the use of local identities over the last decade, our interviewees provided detailed insights in how different local identity discourses were used and adapted by different local stakeholders. It showed how in changing circumstances different elements of identity discourses were played out and aligned to others, generating sometimes new, but mostly rearranged identity discourses. This showed in detail that specific identity discourses are not fixed, but are better conceptualised as social constructs produced and reproduced in discourse, as was discussed in general terms in §2.1 and §2.2. Local identities are not fixed, but are constructed in relation to multiple other identity discourses of sometimes overlapping spaces (§2.7).

This book showed the complex interrelations between identities, places, people and power. Our interviewees link local identities primarily to community values, as was discussed in §4.1. The perceived qualities of the local communities to which one belongs are important elements in the construction of local identity discourses. The key characteristics of local communities are widely shared, but can be appreciated in different ways. This is especially apparent when local identities are linked to future spatial developments. The valuation of heritage sites, but especially building projects are debated with arguments derived from local identity discourses. Sometimes these are based on the fear of the future and the sorrow over the loss of a better past. This was an important element in the opposition in Katwijk against urbanisation. Sometimes they are linked to the hope for a better future to redress the problems rooted in the past. Then the focus is not to oppose new developments but to foster, for instance, regional development or landscape restructuring. On Goeree-Overflakkee, this was linked to the promotion of a new regional identity.

Local identities are most visibly and vocally used in politics to oppose new developments. But identities are used in many different ways. Local and other identity discourses are also frequently linked to future-oriented policies. These tend to be more hidden from public view and are used, for instance, in policy discussions and documents. These are mostly taken for granted by the general public. Policy debates are mostly conducted within the broad framework of accepted spatial identity discourses. Only conflicts affecting key aspects of the different visions on spatial identities bring identity issues into the political debate. This book has discussed many examples of this; on major policy debates on urbanisation (§4.8, §6.1, §6.2), amalgamation (§2.8, §5.1, §6.3), regional development (§5.3, §5.4) and Sunday rest (§5.5), spatial identities and the different views burst into the open. Different visions of the desired future are then frequently linked to expressions of identity. Different views on the future are important driving forces behind the conflicts between (central) administrators and local populations. During municipal mergers especially, the feared vanishing of local identities is frequently used by sections of the population to oppose, in their view, the imposed administrative Leviathan of the amalgamated municipality. The gap between population and administrators is of course a perennial problem, as was discussed in §2.5. Representative democracy is an important method to bridge this gap. Amalgamations can, however, not simply be legitimised by elections for a representative body for the population of the newly formed territory. This argument is frequently used by proponents of amalgamation. However, territorial stability is an important but mostly hidden precondition for an effective representative democracy (§2.3). Changing the borders or powers of territorial administrations changes the political debates away from accommodating differences within the community to articulating differences between communities. When territories are questioned, the very core of what constitutes the community is undermined, which frequently generates passionate conflicts. The feared loss of identity and autonomy fuels conflicts and can generate resistance identity discourses during amalgamations (§2.8, §4.9, §5.1) and a local political system dominated by distributional conflicts between local communities after amalgamation (§6.3).

In other situations, the lack of support and a shared secondary identity shielding the primary local identity of the local community can also generate support for a re-territorialisation. When the gap between population and administration (§2.5) is already wide and the spatial identities and administrative organisation and borders are not in alignment, amalgamation can have wide support because of the expected more efficient and effective administration, controlled by a better equipped representative body. This line of argument was supported by large sections of the population living in the central and eastern parts of Goeree-Overflakkee (§5.3, §5.4).

7.1 The changing use of local and regional identities compared

In both case studies, local identities were a major factor in the municipal amalgamation process. On Goeree-Overflakkee, it was initially dominated by the opposition from the municipality of Goedereede on the western side of the island. Through this conflict on amalgamation, they developed a thickening resistance identity discourse (§2.8, §5.1). Many members of the local communities in Ouddorp and Goedereede feared losing their own effective and efficient municipal administration with a well-established secondary identity of dealing with different views on local identity. They feared losing control over their future when amalgamated with the different municipalities on the rest of the island. A clear resistance identity discourse emerged which focussed on thick identity elements and which was widely supported by the local population.

After amalgamation, this resistance identity discourse diminished and was surpassed by a thinner regional identity discourse focussing on collectively resolving the pressing needs of the island for the future (§5.2). Thus while the resistance identity discourse focussed on the feared future of losing local autonomy, the new regional identity discourse focussed on the future feared especially by leading members of the public and private sector of the island as a whole (§4.10). Public services, like hospitals, and the employment and profitability of the private sector dominated by traditional tourism, light industry and agriculture are all underperforming. The island as a whole is deemed to profit from closer relations with neighbouring cities (§5.3). The dominant identity discourse thus shifted from local protection and resisting interference by neighbouring communities, to regional cooperation in order to better position the island as a whole in its urban environment. The focus of local identities changed from the specific, mostly historically rooted, elements which make local communities on the island different, to the more general elements which they share. These local elements, like the coastal landscape and solidarity, came together with thinner and more general elements like sustainability, part of a new thinner regional identity discourse. The selective downloading of elements like sustainability and the uploading elements from local identities like resilience and solidarity created a layered identity discourse for the island (§2.7). This is an emerging secondary identity, but is not yet developed enough to prevent the use of thicker primary identities in conflicts over Sunday rest (§5.5). In this respect, there are identity gaps (§2.5) between the thin regional identity promoted by the administration and the local entrepreneurs and the thick local identities which are more important for the local population.

In the case of Katwijk, local communities and their identities were not threatened so much by amalgamation, like on Goeree-Overflakkee, but by the collective threat of urbanisation. Initially the threat of a large new town was so big that it resulted in a regional response allying not only the many affected municipalities, but also regional business associations, environmental associations and the province of Zuid-Holland (§6.1). A new regional identity was formulated through combining thick and thin identity elements (§2.4) in a layered new regional identity discourse. For instance, it downloaded the renewed focus of Dutch national politics on competitiveness and uploaded the success stories of local bulb traders (§2.7).

This emergence of a thinner regional identity discourse focussing on regional development is very similar to Goeree-Overflakkee. In both cases, a perceived shared interest and need for collective action was crucial. However, for Katwijk and the Bollenstreek this was more an external threat to regional strength, while on Goeree-Overflakkee the regional identity discourse was more linked to the internal weaknesses of the region and external opportunities. For Katwijk, the success of this regional identity discourse in averting large-scale urbanisation in the Bollenstreek eventually led to its decline (§6.1.4). The later urbanisation threat in Valkenburg was more limited, which generated a more localised response through the voluntary amalgamation of just three of the municipalities in the Bollenstreek into the municipality of Katwijk (§6.2). They not only collectively protected their local identities against external urbanisation, but also internally protected their local identities against each other. This resulted in a local political system based on the jealously guarded fair distribution of services and investment over the different local communities (§6.3). Neighbourhood councils are the most visible aspects of the institutionalisation of this divisive local political system (§6.4). The role of the four different nationalistic Orange associations in the institutionalisation of these differences is, however, far more important (§6.5, §6.6, §6.7, §6.8).

Although local identity discourses focus on the municipality to which they belong, they also position themselves in their wider surroundings. This can be maximising independence from regional cooperation, as in Katwijk, or attracting regional resources, as on Goeree-Overflakkee. Local identity discourses then become linked to expressions of regional identity. On Goeree-Overflakkee, they explicitly use elements from well-established local identity discourses to construct and market their region to the outside world. Local identity discourses are thus primarily rooted in the values attached to local communities, serve as a guide to valuate future developments and focus on the relations with different scales. They are, however, contested and adaptive to changing circumstances.

While on Goeree-Overflakkee local identities were initially used as an opposing force, they later became part of a forward-looking regional identity discourse. Katwijk experienced a very different development of the use of local identities. The very success of their regional identity discourse against external urban and regional threats led in the end to its demise, partly because it became ever more outward looking. Now the continued protection of the different local identities has become part of a very divisive and envy-based, inward-looking secondary identity discourse. Whereas on Goeree-Overflakkee the formulation of an overarching regional identity is seen by many as supporting local identities, in Katwijk the formulation of such an overlaying identity is seen as a threat to the local identities.

7.2 The fragile shield of secondary identities protecting primary identities

This diverse and changing use of local identities during municipal amalgamations can be interpreted using the distinction between primary and secondary identities, which was introduced in §2.9. Primary local identities are relatively stable and well known by the local population and they differentiate between communities. Within local communities, there is widespread cognitive agreement about the character of the local identity, but different opinions on how to affectively value this and how to act upon this in their daily life and in local politics. Within a municipality with multiple local communities, there emerge over time informal ways to accommodate these differences. This specific way of dealing with differences related to local identities becomes a kind of secondary local identity. Local identity discourses not only focus on the primary local identities, but also on the secondary identity of how to deal with these different valuations of primary local identities in daily life and in municipal politics. While primary identities are based on distinctiveness and differences, secondary identities are based on dealing with differences. In §5.5, the analysis of the conflicts on Sunday rest in the new municipality of Goeree-Overflakkee showed how the destruction of secondary identities through amalgamation generates political struggles based on the different valuation (affection) of the different local identities (cognition). This results in conflicts over the regulation of behaviour (conation) in everyday life.

Secondary identities, like any other types of identities, make sense of the strained and changeable relations between elements and collectives. But whereas primary identities focus more on characteristics of separate local communities, secondary identities focus on the relations between communities. Primary identities are thus more place-based, while secondary identities are more relational, although their relations are mostly institutionalised within a political territory. Changes in the power balance within and between local communities and external influences, like amalgamations, make secondary identities less stable than primary local identities based on community characteristics. Whereas primary local identities focus on the specific characteristics of local communities and their differences from one another, secondary identities are in contrast based more on accommodating differences, more outward looking and more focussed on the development of shared interests. Primary identities are thicker and secondary identities are thinner (§2.4). During a municipal amalgamation, secondary identities disappear with the old municipalities. Municipal amalgamations undermine the distinct ways in which communities have learned to live with different valuations of local identities. The disappearance of these secondary identities leaves these primary local identities unprotected. Municipal amalgamations then appear to threaten local communities and their identity. The lifting of the protective shield of a secondary identity exposes the vulnerability of the underlying primary local identities. This propels local identities into the political debate.

7.3 Three different archetypes of using local identities

The detailed discussion on the changing use of local identity discourses in this book has shown how specific local and regional identity discourses were used in different local political situations. The concepts discussed in Chapter 2 helped to clarify this. Thick and thin spatial identities (§2.4), identity gaps (§2.5), the institutionalisation of different spaces (§2.6) with sometimes layered spatial identities (§2.7) and sometimes resulting in the formation of resistance identities (§2.8), culminating in the analysis of the changing relation between primary and secondary identities (§2.9, §7.2) helped to better understand the different ways in which local and regional identities are used, especially during municipal amalgamations. Based on the empirical analyses and these theoretical concepts, three basic archetypes of the use of local identities can be distinguished: local resistance identities, new regional identities and divisive local identities. Thickening local resistance identity discourses are used to oppose amalgamation. Thinner regional identity discourses can bridge local differences to cooperate for development. Different local identities can also be institutionalised within an amalgamated municipality to protect the different local identities, both collectively against the outside world and individually against each other. In the following sections, these different archetypes are further explained. Although mainly based on the analysis of the case studies, these three archetypes are presented here in a more abstract way; they are more ideal types than complex actual practices.

7.3.1 Local resistance identities

In the first archetype, local identities become more inward oriented, focus more on their historical roots and their differences with others; they "thicken" into resistance identities. Identity discourses can have different visions of the future. The feared loss of control through amalgamation frequently results in the development of local resistance identities. Local identity discourses then focus on differences and conflicts. Local resistance identities are used to oppose administrative restructuring like amalgamations.

This fits quite well with how the opposition against amalgamation in the old municipality of Goedereede used their local identity.

7.3.2 New regional identities

However, local identities can also be used to generate support to collectively work for a better future through alignments with similar or complementary identities. Whereas in the first archetype identities thicken, in this second archetype, a thinner identity layer is formulated using both elements of existing thicker, mostly local identity elements and thinner, more regional and general identity elements.

Sometimes a political entity such as a municipality and its secondary identity discourse is too weak to support primary spatial identities. Amalgamation or cooperation can then help to create not only a more effective and efficient organisation, but also a new and more attractive secondary identity discourse. This is a thinner and more regional identity discourse which can be based on a selection of characteristics of established primary local identities and linked to more general policy discourses.

On the island of Goeree-Overflakkee, we saw that local entrepreneurs were instrumental in the formulation of a new thinner regional identity discourse, promoting the island to the outside world and protecting and promoting local identities and interests. This was to a lesser degree also the case in the Bollenstreek in the beginning of this century.

7.3.3 Divisive local identities

A third archetype of how local identities are used in local politics focusses on the protection of local identities through combining the forces of different spatial communities. Amalgamations can create a stronger collective shield against the outside world and local politics can internally actively protect the different local identities. In this archetype, identities operate differently at different scales. Externally it collectively defends local identities against a shared exterior threat like urbanisation. Internally it protects local identities through distributive policies. A secondary identity discourse emerges which focusses on the relation between local communities and the distribution of resources over the local communities.

In Katwijk, this divisive use of local identities is clearly recognisable. The internal distributive focus of the secondary identity discourse was combined with the protection against external urban influences to protect the rural character of its villages.

There are thus many different ways in which local identities can be used in politics to negotiate the old and the new.

References

Delanty, G. & C. Rumford (2005). *Rethinking Europe: social theory and the implications of Europeanization*. London: Routledge.

Elias, N. & Scotson, J.L. (1965). *The established and the outsiders: a sociological enquiry into community problems*. London: Frank Cass.

Paasi, A. (1996). *Territories, boundaries and consciousness: the changing geographies of the Finnish–Russian border*. Chichester: Wiley.

Said, E. (1978). *Orientalism*. New York: Pantheon.

Index

Page numbers in *italics* denote tables, those in **bold** denote figures.

Taylor & Francis eBooks

Helping you to choose the right eBooks for your Library

Add Routledge titles to your library's digital collection today. Taylor and Francis ebooks contains over 50,000 titles in the Humanities, Social Sciences, Behavioural Sciences, Built Environment and Law.

Choose from a range of subject packages or create your own!

Benefits for you

» Free MARC records
» COUNTER-compliant usage statistics
» Flexible purchase and pricing options
» All titles DRM-free.

Benefits for your user

» Off-site, anytime access via Athens or referring URL
» Print or copy pages or chapters
» Full content search
» Bookmark, highlight and annotate text
» Access to thousands of pages of quality research at the click of a button.

REQUEST YOUR **FREE** INSTITUTIONAL TRIAL TODAY

Free Trials Available
We offer free trials to qualifying academic, corporate and government customers.

eCollections – Choose from over 30 subject eCollections, including:

Archaeology	Language Learning
Architecture	Law
Asian Studies	Literature
Business & Management	Media & Communication
Classical Studies	Middle East Studies
Construction	Music
Creative & Media Arts	Philosophy
Criminology & Criminal Justice	Planning
Economics	Politics
Education	Psychology & Mental Health
Energy	Religion
Engineering	Security
English Language & Linguistics	Social Work
Environment & Sustainability	Sociology
Geography	Sport
Health Studies	Theatre & Performance
History	Tourism, Hospitality & Events

For more information, pricing enquiries or to order a free trial, please contact your local sales team: www.tandfebooks.com/page/sales

Routledge
Taylor & Francis Group

The home of Routledge books

www.tandfebooks.com

Printed in the United States
by Baker & Taylor Publisher Services